THE MONTE CARLO METHOD
ENGINEERING APPLICATIONS

M. Mazdrakov   D. Benov   N. Valkanov

ACMO
ACADEMIC PRESS

Sofia, Bulgaria, 2018

Reviewers: Prof. **Radi Radev** (St. Ivan Rilski University of Mining and Geology, Sofia), Prof. **Violin Karagiozov,** PhD (American University in Bulgaria, Blagoevgrad)

ACMO Academic Press
81V Bulgaria Blvd., 1404 Sofia, Bulgaria
+359 (878) 358-647, www.acmo-pub.com

Interior design and preparation by **ACMO Academic Press Ltd.**, acmo-pub.com
Cover by **St. Georgiev** and **K. Iliev** – "Wonder Swamp" Ltd., wonderswamp.com
Printed by **Direct Services Ltd.**, directservices.bg
Translated from Bulgarian by **Lozanova 48 Ltd.**, lozanova48.com

0 9 8 7 6 5 4 3 2 1

ISBN 978-619-90684-3-4 (English Edition)
ISBN 978-619-90684-1-0 (Bulgarian Edition)

Series **Science/Practice**
Series Editors **M. Mazhdrakov, D. Benov, D. Benova**

The following software products are used in this monograph
CADMin ACMO and **Urban Acoustics** (M. Mazdrakov, D. Benov etc., Benovi Engineering Ltd., www.benov.org), **GeoSlope** (Geo-Slope Intl. Ltd., www.geo-slope.com), **Guesstimate** (getguesstimate.com).
This monograph contains material that is copyright of third parties
Figure II-2. Excerpts from the Table of Random Digits of The RAND Corporation (1955) © **RAND Corporation, Santa Monica, CA,** http://www.rand.org/pubs/monograph_reports/MR1418.html.
Point B.V.4. Determining of stability coefficient. © **GEO-SLOPE International Ltd., Calgary, Alberta, Canada,** www.geo-slope.com.

THE MONTE CARLO METHOD
ENGINEERING APPLICATIONS

M. Mazdrakov   D. Benov   N. Valkanov

ACMO
ACADEMIC PRESS

Sofia, Bulgaria, 2018

Editors

**Prof. DSc.
Metodi Mazhdrakov**
St. Ivan Rilski
University Of Mining
And Geology, Sofia

Dobriyan Benov
Benovi Engineering Ltd
Sofia, Bulgaria
benov@benov.org

**Prof. DSc.
Nikolai Valkanov**
Minstroy Holding AD
Sofia, Bulgaria
office@minstroy.com

Contributors

Prof. DSc. Nikolai Nikolov
Institute of Building Physics, Technology and
Logistics AD
Sofia, Bulgaria
dtnnikolov@abv.bg

Assoc. Prof. Georgi Trapov, PhD
St. Ivan Rilski University Of Mining And
Geology
Sofia, Bulgaria
trapov@abv.bg

Prof. DSc. Stoyan Hristov
St. Ivan Rilski University Of Mining And
Geology
Sofia, Bulgaria

Prof. Koino Yovev, PhD
St. Ivan Rilski University Of Mining And
Geology
Sofia, Bulgaria
koboev@abv.bg

Assoc. Prof. Stanislav Topalov, PhD
St. Ivan Rilski University Of Mining And
Geology, Sofia, Bulgaria
stopalov@gmail.com

Diyana Benova
Benovi Engineering Ltd
Sofia, Bulgaria
benova@benov.org

Prof. Ilinka Ivanova, PhD
Konstantin Preslavsky University of Shumen
Shumen, Bulgaria
ilinkaivanova@mail.bg

Mitko Dimov, PhD
St. Ivan Rilski University Of Mining And
Geology
Sofia, Bulgaria
mit.eri@abv.bg

Assoc. Prof. Veselin Hristov, PhD
St. Ivan Rilski University Of Mining And
Geology
Sofia, Bulgaria
veso@mgu.bg

Ivaylo Ivanov
Ellatzite Mine Complex
Etropole, Bulgaria
i.g.ivanov@ellatzite-med.com

Ivan Vasilev
Ellatzite Mine Complex
Etropole, Bulgaria
i.vasilev@ellatzite-med.com

Eliza Mitova
University of Zürich
Zürich, Switzerland
eliza.mitova@uzh.ch

Vasil Shishkov
Ellatzite Mine Complex
Etropole, Bulgaria
v.shishkov@ellatzite-med.com

Stoyo Bosnev
Minproekt EAD
Sofia, Bulgaria
bosnev@minproekt.com

English
language editor

Georgi Petrov
Ellatzite Mine Complex
Etropole, Bulgaria
g.petrov@ellatzite-med.com

TABLE OF CONTENTS

PREFACE

The barriers between scientific disciplines, departments, research units and institutes have been repeatedly identified as a major drawback/obstacle to the development of scientific knowledge and to the use of scientific achievements to solve different tasks, including practical. The overcoming of these barriers and the transition to multi- and interdisciplinary research is a complex and long-lasting process that is actively stimulated by the needs of social practice.

This monograph is a significant step in this difficult and continuous process. The conception of the monograph is well-formulated. The high scientific level is provided by the competence of the scientific editors and the authors' team. Their rich and perennial research experience is a significant condition for the high scientific level of the monograph. It is also important that there is an active collaboration between leading researchers, lecturers and business executives. This will stimulate future research and provide the theoretical basis for a variety of management decisions

The dignity of the monograph is that a long historical period – from the appearance of the Manhattan Project (1942-1946) to the present day – has been considered.

The author's team has been successfully selected to meet the high-level requirements for solving a wide range of practical tasks related to the design and management of extraction in underground mines, urban acoustics, modeling of complex processes in a probable environment, etc.

Prof. Radi Radev

TO THE DEAR READER

> *I look to the left... NOTHING?*
> *I look to the right... NOTHING?*
> *So, I say to myself: There is SOMETHING here...*
>
> —— *From Burgas folklore sayings about*
> *Uncle Petyo Banderata.[1]*

One of mankind's successful attempts to find out what that SOMETHING is the Monte Carlo Method.

The method, as well as many of the achievements of mankind, was created for military purposes as part of the scientific tasks associated with the creation of the atomic bomb. The event was super secret and everything was encrypted. The code name of the method – Monte Carlo, has proved to be very successful and has survived in civilization (suck fate has the name of the armoured fighting vehicle – tank).

The task was to create a method for modeling the behavior of a complex probability system. The classic solution is to present the phenomenon with one, two, etc. (but always a limited number) indicators. The new solution is the opposite – "artificially" increasing the number of input/output information.

Currently, the Monte Carlo Method is effective, and in some cases – the only one, solution for a wide range of tasks from all areas of scientific knowledge. That is why we've decided to present yet

---

1 Legendary Bulgarian kidder

another exposure of the foundations and some of the Monte Carlo applications.

The monograph is divided in two parts.

The first part returns the reader during the World War II. We follow the development of the idea of the method and the associated need for creating a powerful enough computer. The first publications are mentioned and are examined the scientific basics of the method and some basic algorithms.

Described is application of the method for solving the classical task – calculating the number π and for two malicious problems – the phenomenon Black Swan and the functional literacy of the students.

The second part contains applications of Monte Carlo method for solving tasks that can be characterized as "engineering". Without neglecting the concrete results obtained, we will point out that the described approaches for the practical application of the Monte Carlo method are of the greatest interest.

The monograph includes tasks from the following areas:

– quality management in extraction of underground resources – nonlinear optimization task solved in complex and dynamic conditions;

– stability of opencast slopes in complex and underdeveloped environments;

– assessment of the accuracy of mine surveying and geodetic measurements;

– productivity of machinery in open pit mines;

– emissions modeling;

– urban acoustics.

The present monograph is an attempt to examine the basic features of the Monte Carlo method, to systematize the results and conclusions of its application by authors and other researchers and, as a result, to help others using the method for solving a wide range of tasks in various engineering sciences.

The authors believe that the monograph is of interest to specialists who work in the exploration and extraction of underground resources, environmental protection, urban planning and construction, mathematical modeling, and application development, engineering geology and geomechanics. We hope that the monograph will also be useful for students and PhD students from a significant number of professional fields.

But the most useful monograph will be for those who are convinced that the world is probable and that's why life is so interesting.

Sofia, 2018

M. Mazhdrakov, D. Benov, N. Valkanov

AKNOWLEDGEMENTS

The editors would like to thank Minstroy Holding AD, Sofia for the invaluable help during the writing of the monograph.

We would also like to thank RAND Corporation, Santa Monica, CA, and GEO-SLOPE International Ltd., Calgary, Alberta, for the kind permission to use their copyrighted materials during the preparation of this book.

We would also like to thank our co-authors who have enriched the contents of the monograph.

We thank Aleksadar Tsonkov who directed our attention to the Black Swan problem.

Last but not least we would also like to express our gratitude to the team that worked on the presentation and outlook of this book.

PART 1

INTRODUCTION:
the method

> " As I have said so many times,
> God doesn't play dice with the world. "
>
> **Albert Einstein** (1943)

JOURNAL OF THE AMERICAN STATISTICAL ASSOCIATION

SEPTEMBER 1949

THE MONTE CARLO METHOD

NICHOLAS METROPOLIS AND S.

Los Alamos Lab

Volume 44

SECTION I.
THE BEGINING

D. Benov, M. Mazdrakov, N. Valkanov

The story of the Monte Carlo method is linked to two very important events in the modern history of mankind: the World War II (the Manhattan Project, in particular) and the creation of the first electronic computers. For this reason, we will try to synthesize from these events the most important facts that are relevant to the creation of the Monte Carlo method of stochastic modelling in order to get a fuller picture of its emergence (R. Eckhardt, 1987; D. Benov, 2016).

One of the creators of the method, Nicholas Metropolis, opens his article *"The Beginning of Monte Carlo"* (1987) with the following words:

> *The year was 1945. Two earthshaking events took place: the successful test at Alamogordo and the building of the first electronic computer.*
>
> — *Nicholas Metropolis (1987)*

I.1. THE MANHATTAN PROJECT (1942-1946)

On August 2nd, 1939, The Einstein – Szilárd letter written by Leó Szilárd after consulting his colleagues, the Hungarian physicists Edward Teller and Eugene Winger, and signed by Albert Einstein was sent to US President Franklin D. Roosevelt. The letter stated that according to the new research the element of uranium could be subjected to nuclear fission with subsequent release of a tremendous quantity of energy. It also stated the likelihood of causing a nuclear reaction and the possibility of making exceptionally powerful bombs. The authors pointed out that such bombs could be transported by ship but were skeptical about their transportation by airplane. Einstein noted that most likely Germany was also conducting similar scientific research, because the Nazis had stopped the export of uranium ore from the mines in the occupied Czech territories, Roosevelt was urged to take measures being the start of the Manhattan Project.

The Manhattan Project (L. Groves, 1962; S. Groueff, 1967) is the code name of the project to produce the first nuclear weapons in the world. It was carried out in the USA with the assistance of the United Kingdom during the World War II. The name comes from the initial headquarters of the project: the borough of Manhattan in New York City. Initially, the project was called "Manhattan Engineer District" but subsequently everybody referred to it as simply "Manhattan".

All operations were carried out under the control of the US Army Corps of Engineers and the project itself was under the direction of Gen. Leslie Groves. The conceptual and scientific director and mastermind of the Project was J. Robert Oppenheimer, a scientist of world renown. In the project took part many Nobel prize winners such as Niels Bohr, Enrico Fermi, Richard Feynman, Eugene Winger, Hans Bethe, Glenn Seaborg, Luis Alvarez and other scientists and nuclear specialists of world renown. The average age of the team was 26 years.

As a result of these efforts in 1945 three nuclear bombs were built and detonated: two of them are plutonium ones – "Fat Man" (named in honor of Winston Churchill) and "Trinity", and the third one is the uranium bomb "Little Boy" (a reference to former President Roosevelt).

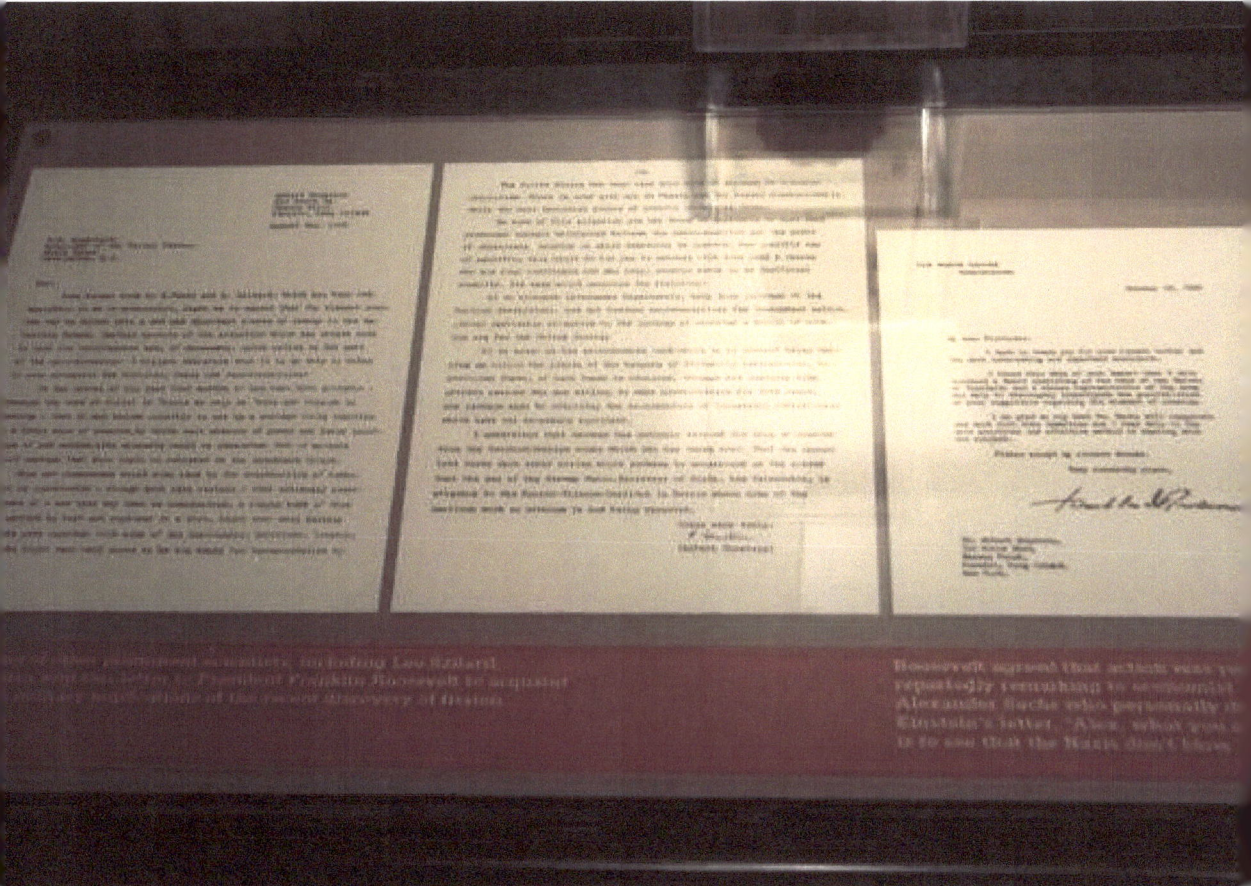

Figure I-1. Einstein – Szilárd letter and Franklin D. Roosevelt's response on display at Bradbury Science Museum, Los Alamos

President F. D. Roosevelt signing the order launching the Manhattan Project (May 12, 1942) (archives of National Nuclear Security Administration)

"Trinity" was the first bomb to be detonated: on July 16, 1945, in the test area near Alamogordo, New Mexico. This is the first man-made nuclear test. It was a tremendous success far exceeding the expectations.

The exact origin of the code name "Trinity" for the test is unknown, but it is often attributed to Oppenheimer as a reference to the poetry of John Donne, which in turn references the Christian notion of the Trinity (three-fold nature of God) (L. Groves, 1962).

On August 6, 1945, President Truman ordered that the "Little Boy" be dropped above the Japanese port of Hiroshima from the bomber Boeing B-29, and on August 9 the same year "Fat Man" was dropped above Nagasaki. No nuclear weapon has ever been used against civilian population since then.

> *Japan was also on its way to produce a atomic bomb, a fact that is not very popular. The Japanese tried to get an atomic bomb and in the years of the World War II they did extensive work on that technology. When in December 1938, the German physicists Otto Hahn and Fritz Strassmann discovered the nuclear fissiono, physicists form the leading industrial states quickly realized the military dimensions of that discovery. The Japanese researcher Yoshio Nishina (1890-1951), who was a good friend of Niels Bohr and Albert Einstein, appreciated the discovery's military application as early as 1939. Two years later, in April 1941, Nishina was officially commissioned to study the possibilities of making an atomic bomb. However, the n project did not produced the expected outcome and, in the meantime, it was established that the Japanese uranium mines could not provide the 21 tons of enriched uranium necessary for the bomb. Subsequently, the laboratory was destroyed by US bombers and after the end of the war no attempts have been made to restore it. So the N project, named after the first letter of its director Nishina's name, finally failed (E. Lilov, 2013).*

The production of necessary materials was kept in strict secrecy and was spread in dozens of places all over the United States (Figure I-5). The most important units of the project were the uranium enrichment facility in Oak Ridge, Tennessee, and the laboratory in Los Alamos, New Mexico. Approximately 129 thousand people in total were engaged in the project as most of them did not know what the purpose of their work was.

Albert Einstein (1879-1955) German theoretical physicist, philosopher and writer. Nobel-prize winner

Gen. Leslie Richard Groves (1896-1970) General of US Army Corps of Engineers

Leo Szilárd (1898-1964) American physicist, one of the creators of the concept of nuclear chain reaction

Enrico Fermi (1901-1954) Italian physicist. Nobel-prize winner

Eugene Wigner (1902-1995) American physicist and mathematician. Known as the "Silent Genius". Nobel-prize winner

John von Neumann (1903-1957) American mathematician

J. Robert Oppenheimer (1904-1967) American theoretical physicist. Known as the "father of atomic bomb"

Edward Teller (1908-2003) American theoretical physicist, one of the creators of the hydrogen bomb

Stanisław "Stan" Ulam (1909-1984) Polish-born American mathematician

Robert Richtmyer (1910-2003) American physicist and mathematician, lecturer

Nicholas Metropolis (1915-1999) American mathematician and physicist

Andrei Dmitrievich Sakharov (1921-1989) Soviet nuclear physicist. One of the creators of hydrogen bomb

Figure I-2. J. Robert Oppenheimer and Gen. Leslie Groves at the test site in Alamogordo after the successful Trinity test

Figure I-3. Statues of Gen. Leslie Groves and J. Robert Oppenheimer at Bradbury Science Museum, Los Alamos (Ron Cogswell, Flicr)

"All the News That's Fit to Print"

The New York Times.

LATE CITY EDITION

FIRST ATOMIC BOMB DROPPED ON JAPAN; MISSILE IS EQUAL TO 20,000 TONS OF TNT; TRUMAN WARNS FOE OF A 'RAIN OF RUIN'

HIRAM W. JOHNSON, REPUBLICAN DEAN IN THE SENATE, DIES

Jet Plane Explosion Kills Major Bong, Top U. S. Ace

KYUSHU CITY RAZED

Kenney's Planes Blast Tarumizu in Record Blow From Okinawa

ROCKET SITE IS SEEN

125 B-29's Hit Japan's Toyokawa Naval Arsenal in Demolition Strike

REPORT BY BRITAIN

'By God's Mercy' We Beat Nazis to Bomb, Churchill Says

ROOSEVELT AID CITED

Raiders Wrecked Norse Laboratory in Race for Key to Victory

Steel Tower 'Vaporized' In Trial of Mighty Bomb

NEW AGE USHERED

Day of Atomic Energy Hailed by President, Revealing Weapon

HIROSHIMA IS TARGET

'Impenetrable' Cloud of Dust Hides City After Single Bomb Strikes

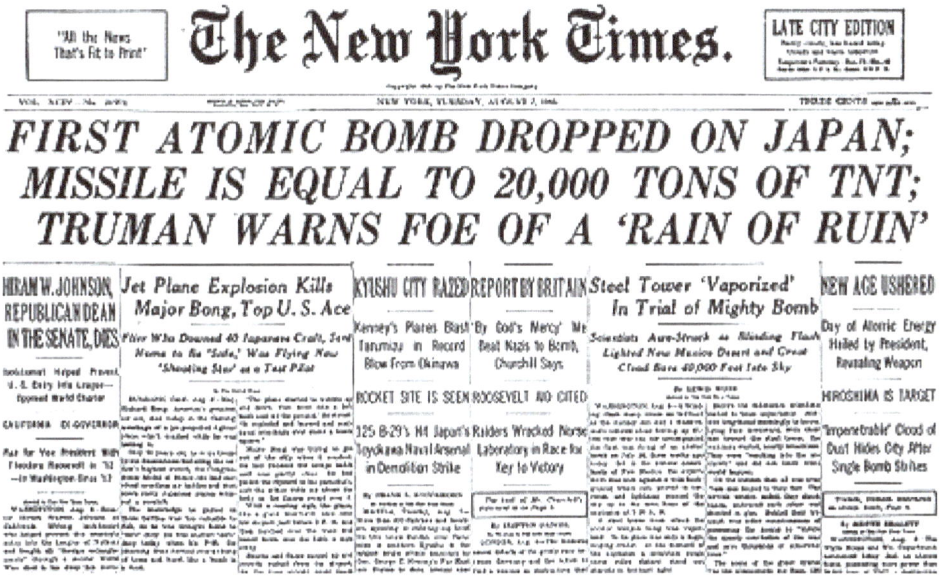

Figure I-4. The front page of The New York Times on August 6, 1945 (source: The New York Times: On This Day)

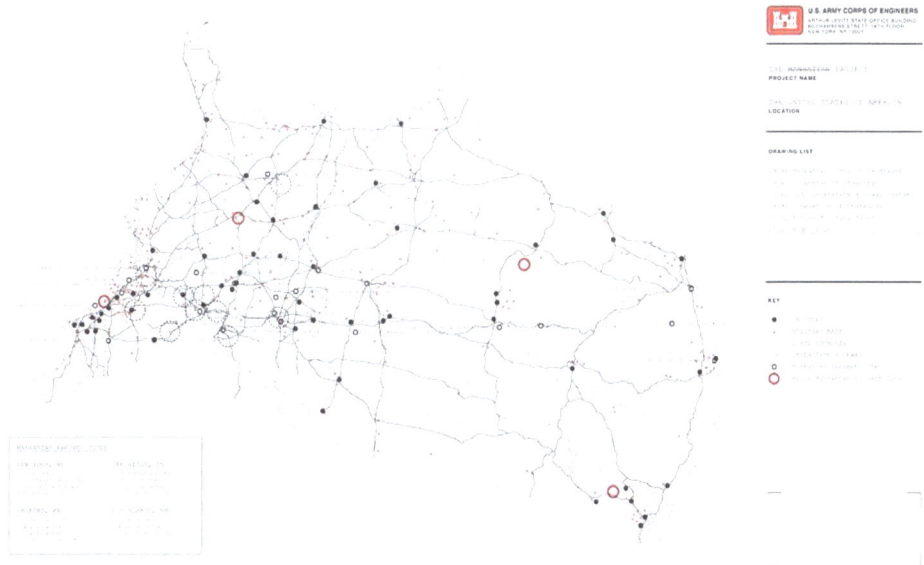

Figure I-5. Map of Manhattan Project locations

The Manhattan Project was an absolute priority of US government. The financial resources were virtually unlimited. The project started with a 6,000-dollar budget. Five years later, however, the sum rose to $2 billion (more than $30 billion in 2010 after an inflation adjustment!). According to the director of Los Alamos Historical Museum: "*This is the second most expensive American research project after the Moon Landing mission*" (G. Schließ, 2015).

The successful end of the project was marked on July 18, 1945.

On November 10, 2015, the official opening of "*Manhattan Project National Historical Park*" took place. It consists of 3 sites: Oak Ridge (TN), Hanfrod (WA), and Los Alamos (NM).

A curious fact is that according to Forbes magazine data as at 2012 (N. Vardi), Los Alamos is the richest town in the State of New Mexico, and the fourth richest county in the USA.

Figure I-6. July 18,1945, Cecilienhof: [From left to right, first row]: Josef V. Stalin, Harry S. Truman, Soviet Ambassador to the United States Andrei Gromyko, Secretary of State James F. Byrnes and Soviet Foreign Minister Vyacheslav Molotov. Second row: general Harry Vaughan, Russian interpreter Charles Bohlen, and President's naval aide James Vardaman.

Figure I-7. November 10, 2015: Manhattan Project Nation Park is established

Figure I-8. Today, the place where the first atomic bomb test was conducted (Alamogordo, New Mexico), is open for visitors (Photograph: Gero Schließ, Deutsche Welle)

1.2. THE IDEA

Enrico Fermi used a probabilistic approach in the calculation of the neutron diffusion but did not publish details about the method used (H. Anderson, 1986b).

In 1946, the physicists from Los Alamos working under the Manhattan Project studied the distance likely to be traveled by the neutrons in different materials (G. King, 1951). Although most of necessary data were correctly calculated Los Alamos physicists were unable to solve the problem with the help of conventional, determined mathematical methods.

Then Stanisław Ulam formulated the idea of using random experiments.

The first thoughts and attempts I made to practice [the Monte Carlo method] were suggested by a question which occurred to me in 1946 as I was convalescing from an illness and playing solitaires. The question was what are the chances that a Canfield solitaire laid out with 52 cards will come out successfully? After spending a lot of time trying to estimate them by pure combinatorial calculations, I wondered whether a more practical method than "abstract thinking" might not be to lay it out say one hundred times and simply observe and count the number of successful plays. This was already possible to envisage with the beginning of the new era of fast computers, and I immediately thought of problems of neutron diffusion and other questions of mathematical physics, and more generally how to change processes described by certain differential equations into an equivalent form interpretable as a succession of random operations. Later... (in 1946) described the idea to John von Neumann and we began to plan actual calculations.

— *Stan Ulam (1991)*

Ulam proposed the Monte Carlo method for analysis of complex mathematical formulae related to the theory of nuclear reactions. This proposal led to a more systematized development of the method by von Neumann, Metropolis and others, which considerably aided the resolving of many of the most complex problems in the making of the atomic bomb (T. Rajchel, 2013).

The Monte Carlo method also played a main part in the simulations made after the end of the World War II during the sequel to the Manhattan Project to make a hydrogen bomb, also called thermonuclear weapon (K. W. Ford, 2015), or weapon based on Teller-Ulam design (L. Ackland, 1990; J. Richelson, 2007). In the Soviet Union, the term *"Andrei Sakharov's Third Idea"* was popular (G. Gorelik, 2011).

In 1950, the Monte Carlo method was used in Los Alamos in the development of the hydrogen bomb and became popular in the developments in the field of physics, physical chemistry and operations research (R. Richtmyer, J. von Neumann, S. Ulam, 1947). RAND [2] Corporation and US Air Force played a major part in financing and spreading the Monte Carlo method in many different fields.

RAND Corporation, founded in 1948 and stil existing today, is the heir to the Project RAND of United States Army Air Forces (A. Abella, 2008). It is the most famous (the most successful!) US strategic research center. Initially, it engaged in building airplanes, rockets and artificial satellites. Subsequently, it engaged in computing equipment and programming.

Being secret (Figure I-4), the work of von Neumann and Ulam required a code name. N. Metropolis (1987) suggested the name "Monte Carlo", an allusion of the casino of the same name (Figure I-10), where Stan Ulam's uncle – Michael Ulam, spent a lot of time.

Though the method has been criticized as crude ("drudgery"), von Neumann justified it as being faster than any other method, and also noted that when it went awry it did so obviously, unlike methods that could be subtly incorrect.

---

2 Research ANd Development

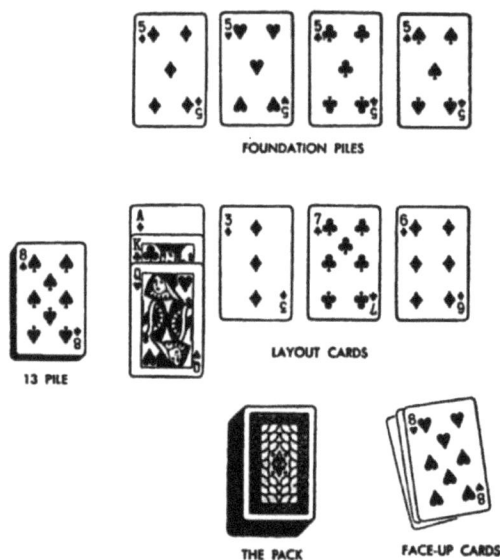

Figure I-9. Layout for Canfield Solitaire (J. Leeming, 1980)

> **"** [*The procedure is*] ... *analogous to the playing of a series of "solitaire" card games and is performed on a computing machine. It requires...* "*random*" *numbers with a given distribution.*
>
> — *Stan Ulam, John von Neumann (1947)*

At that time Ulam and von Neumann did not provide further details in their abstract and there are no traces of a published article. A short time before his death in a conversation with Cuthbert Hurd, Stan Ulam said he did not remember his material and referred him to Los Alamos report by Richtmyer, von Neumann and Ulam (1947).

This report describing a stochastic model of physical processes and giving a procedure for calculation of ENIAC.[3] (A. Burks, 1981; L. De Mol, 2009), is not very popular as well. It was classified as "secret" and was not declassified until July 31, 1959. Only eight copies of that report were printed.

---

3 Electronic Numerical Integrator And Computer

Figure I-10. Monte Carlo casino in Monaco, which gave its name to the method

Figure I-11. Part of the first page of the unclassified report by R. Richtmyer, J. von Neumann, S. Ulam (1947)

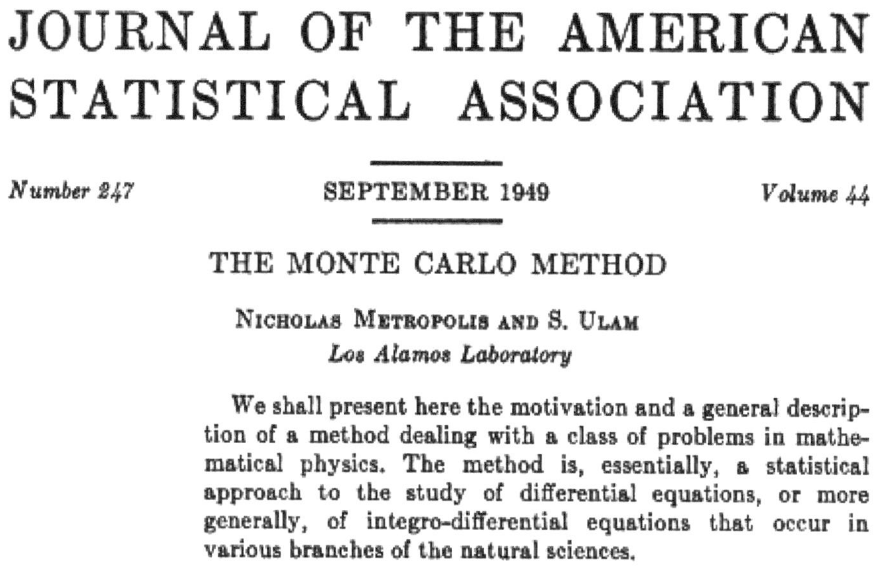

JOURNAL OF THE AMERICAN STATISTICAL ASSOCIATION

Number 247   SEPTEMBER 1949   *Volume 44*

THE MONTE CARLO METHOD

NICHOLAS METROPOLIS AND S. ULAM
Los Alamos Laboratory

> We shall present here the motivation and a general description of a method dealing with a class of problems in mathematical physics. The method is, essentially, a statistical approach to the study of differential equations, or more generally, of integro-differential equations that occur in various branches of the natural sciences.

Figure I-12. Part of the first page of the publication of N. Metropolis, S. Ulam (1949)

I.3.   THE FIRST ELECTRONIC COMPUTERS

It is accepted that the principle of modeling using the Monte Carlo method requires considerable computing resources. Thus, its practical use is linked to the creation of electronic computing machines (N. Metropolis, J. Howlett, G. C. Rota, 1980; N. Metropolis, 2014).

In 1642, Blaise Pascal built a computing machine for two arithmetic operations. He is deemed the construction of the first digital calculator (P. E. Ceruzzi, 2003).

The idea of a programmable computer was conceived by British mathematician Charles Babbage, and the first program was written by Ada King, Countess of Lovelace and daughter of English poet Byron (A. Hyman, 1982; W. Isaacson, 2014).

In 1946, Enrico Fermi made an analog computer to apply the Monte Carlo method in the study of neutron transport in different types of nuclear systems called Fermiac.[4] and also known as "*The Monte Carlo Trolley*" (Los Alamos National Laboratory, 1987).

ENIAC is the first general-purpose electronic computing programmable machine (computer). It was made in 1945 by Americans John Mauchly and John Presper Eckert. The initial purpose was to calculate ballistic tables (tables for artillery fire) for the needs of US Army (W. Isaacson, 2014).

In 1943, the production of ENIAC was assigned to the University of Pennsylvania. The price of development was almost half million dollars. The machine was shut down in November 1946 for a memory refurbishment and was transferred to Aberdeen Proving Ground, Maryland, the next year. It was in use until October 1955. The machine weighed 30 t, occupied an area of 1460 m^2 and consumed 150 kWh of electrical energy.

It should be noted that in 1939 John Atanasoff and Clifford Berry from Iowa State University created an entirely new type of machine built on the basis of electronic elements, which calculated by binary digits and uses memory separated from the processor (in the present-day meaning of the word). This computer was called ABC[5]. Although it is not programmable ABC is considered the first electronic digital computer (J. V. Atanasoff, 1984; W. Isaacson, 2014).

In fact, W. Isaacson (2014) makes the definition of the first modern computers as binary, electronic and programmable.

After the creation of ENIAC, Mauchly and Eckert formed the Eckert–Mauchly Computer Corporation (1947). In 1950, they sold the company to Remington Rand, which, in its turn, amalgamated with Sperry Corporation. The two corporations formed Sperry Rand Corporation, whose Vice President after the end of the war was Gen. Leslie Groves. Today, Sperry Rand Corp. is named UNISYS.

---

4 FERMI Analogue Computer
5 Atanasoff-Berry Computer

The first computers (of ENIAC type), need a physical change of component connection to perform different operations and therefore are referred to as "*fixed-program computers*". As the concept of central processor is commonly linked to the possibility of performing different computer programs the earliest devices that can be called so are the first stored-program computer.

It is accepted that a programmable computer is one where the program instructions are written in electronic computer memory. This definition is often extended by the requirement that program processing and data in memory should be interchangeable and even.

Computers with von Neumann architecture store program instructions and data in the same memory while computers with Harvard architecture have separate memories for programs and data (Figure I-13) (H. G. Cragon, 2000).

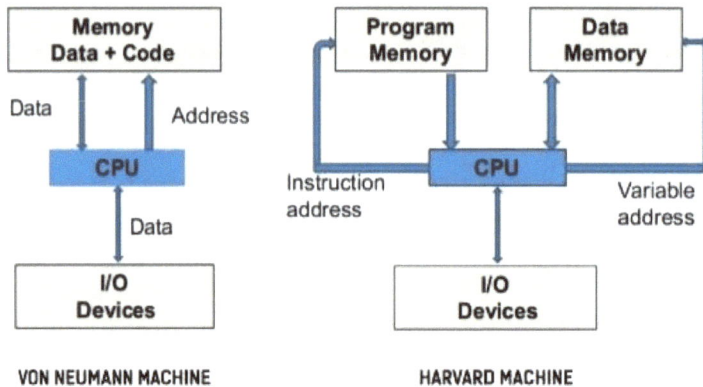

Figure I-13. Scheme of von Neumann and Harvard machines

Sometimes "programmable computer" is used synonymously with von Neumann architecture but others deem it historically incorrect. The first machines with Harvard architecture were treated as "reactionary" by the supporters of programmable computers.

In 1936, the theoretical concept for Turing's universal machine, i.e. programmable computer, was published (A. Turing, 1936; C. Petzold, 2008). Von Neumann was familiar with that work and recommended it to his associates.

Blaise Pascal (1623-1662) French mathematician, physicist, religious philosopher, theologian and writer

Charles Babbage (1791 – 1871) British mathematician, philosopher and inventor

Ada Lovelace (1815-1852) Augusta Ada King-Noel, Countess of Lovelace. British mathematician and writer

Howard Aiken (1900-1973) American physicist and computer pioneer

John Atanasoff (1903-1995) American physicist, mathematician and electrical engineer of Bulgarian origin

Tommy Flowers (1905-1998) British engineer

John Mauchly (1907-1980) American physicist and engineer

Konrad Zuse (1910-1995) German computer specialist

Cuthbert Hurd (1911-1996) American computer engineer and entrepreneur

Alan Mathison Turing (1912-1954) British mathematician, logician, cryptanalyst, computer scientists and philosopher

Nicholas Metropolis (1915-1999) American mathematician and physicist

John Presper Eckert (1919-1995) American electrical engineer and computer pioneer

Many of the first computers, for example, the Atanasoff-Berry computer, are not freely programmable. They perform only one program set by the hardware. For there are no program instructions it is not necessary to record the program in the memory. Other computers, though programmable, store their programs on a punched tape.[6], which is mechanically fed to the computer.

He filed two patent applications in which he anticipated that machine instructions would be stored in the same place where data is stored.

On June 30, 1945, before the completion of ENIAC, John von Neumann published a research paper entitled *"First Draft of a Report on the EDVAC*[7]*"* (1945), in which he described the principles of a computer with stored resident program to be completed in August 1949.

EDVAC was designed for the performance of a certain number of different instructions, which can be combined in order to form applied programs.

The most important novelty of EDVAC is that the programs are stored in high-speed memory instead of being entered by means of physical connection of computer's components. Thus, the hardest limitation of ENIAC – considerable time and efforts necessary to re-configure the computer to perform a new task – was overcome. In von Neumann's computer the program can only be replaced by changing the content of computer's memory. ENIAC had been reconstructed in a way enabling it to store programs even before EDVAC was finished. Unfortunately, ABC was not improved because of the outbreak of the war.

The system IBM ASCC.[8] (also known as Harvard Mark I), suggested on the 1930s by Howard Aiken (a student at Harvard at the time) and finished by IBM before EDVAC by IBM, also uses stored programs but they are recorded on a paper punched tape instead of electronic memory. The main difference between EDVAC and Mark I is that in Mark I data storage and processing is separated from

---

6 The authors have certain experience with fast reading and decoding of punched tape and fully understand the difficulties with the practical application of this technology
7 Electronic Discrete Variable Automatic Computer
8 IBM Automatic Sequence Controlled Calculator

instructions (Harvard architecture), while in von Neumann's system, as in most contemporary processors, they use the same memory (Figure I-13). Aiken calls this machine *"the realized dream of Babbage"* (W. Isaacson, 2014).

In 1936, Konrad Zuse started to work on Z1 – a binary, electromechanical calculator with limited programmability reading instructions from a perforated film. He finished Z1 in 1938. This machine never worked well due to insufficient mechanical precision. Z1 and its blueprints were destroyed during the World War II. In 1941, Zuse built the Z3 machine. It is a binary, 22-bit calculator featuring programmability possibilities, with memory and a calculation unit.

In 1988, Zuse recreated Z1 but suffered a heart attack midway through the project. The machine was made up of 30,000 components, cost 800,000 DM and required four individuals (including Zuse himself) to assemble it. The funding for this project was provided by Siemens and a consortium of five other companies.

In 1943, Tommy Flowers, an engineer in the British Post Office, built the Colossus computer used in Bletchley Park, a secret military laboratory base in Bletchley, Milton Keynes, Buckinghamshire, Great Britain.

In 1949, the EDSAC[9] computer was made in Cambridge based on von Neumann architecture. The team was led by Sir Maurice Wilkes inspired by von Neumann's paper for EDVAC.

During the first calculations under the Monte Carlo method Richtmyer also used IBM SSEC[10] – an electromechanical computer whose design started in 1944. Then appeared IBM CPC[11] (1949), which turned out to be a very popular product. On the basis of that quest Cuthbert Hurd advised new president of IBM Thomas Watson

---

9 Electronic Delay Storage Automatic Calculator
10 IBM Selective Sequence Electronic Calculator
11 IBM Card-Programmed Electronic Calculator

Jr. to build the first IBM commercial stored program computer IBM 701 (1952).

Hurd worked at IBM from 1949 to 1962 where he founded the Applied Science Department and hired John von Neumann as a consultant.

Table I-1. First tasks solved on MANIAC

1952 Fermi, Metropolis *The pion – proton phase shift analysis*

1953 Fermi, Pasta, Ulam Nonlinear coupled oscillators

1954 Bethe, deHoffmann, Metropolis *Phase shift analysis*

1954 Gamow, Metropolis *Genetic code*

1954 Metropolis, Teller *Equation of state: Importance sampling*

1954 Metropolis, Von Neumann *Two-dimensional hydrodynamics*

1954 Metropolis, Turkevich *Nuclear cascades using Monte Carlo (N. Metropolis, R. Bivins, M. Storm и др., 1958)*

1956 Wells *Anti-clerical chess*

1956 Metropolis, Ulam *The lucky numbers*

1973 Metropolis, Stein, Stein *Universalities of iterative functions*

In 1952, the MANIAC [12] I, created under the direction of Nicholas Metropolis was put in use at Los Alamos Laboratory (H. Anderson, 1986a). MANIAC uses von Neuman architecture (Figure I-13). In terms of physics, the digital Universe became a reality in late 1950s in Princeton by building of the MANIAC – at high speed, stored programs, etc. (J. Holt, 2012).

---

[12] Mathematical ANalyzer, Integrator And Computer

In 1961, MANIAC III was built for use at the Institute for Computer Research at the University of Chicago.

On the basis of the computers of the MANIAC series, Metropolis, Teller, von Neumann, Ulam and Richtmyer created algorithms and programs that enabled the development and widespread use of the Monte Carlo method.

Figure I-14. J. Robert Oppenheimer and John von Neumann in front of MANIAC, Princeton, 1952 (Alan Richards/ Emilio Segrè Visual Archives)

In his article "*Scientific Uses of the MANIAC*" Anderson (1986b) gives some interesting examples of the first tasks solved by using MANIAC (Table I-1).

2400 BC Abacus was created in Babylon

300 BC Pingala devised the binary numerical system

60 Hero of Aleksandria devised a machine following a sequence of instructions

1492 Leonardo da Vinci drew a flying machine, the first mechanical calculator and one of the first programmable robots

1642 Blaise Pascal built a computing machine for two arithmetical operations

1822 British mathematician Charles Babbage voiced the idea of a programmable computer, and the first program was written by Ada King, Countess of Lovelace

1936 The theoretical concept for Turing's universal machine was published

1930s IBM ASCC (Automatic Sequence Controlled Calculator) / Harvard Mark I

1936 Z1 (Zuse-1)

1939 ABC (Atanasoff-Berry Computer)

1941 Z3 (Zuse-3)

1943 Colossus

1944 IBM SSEC (Selective Sequence Electronic Calculator)

1945 ENIAC (Electronic Numerical Integrator And Computer)

1945 EDVAC (Electronic Discrete Variable Automatic Computer)

1946 FERMIAC (FERMI Analogue Computer)

1949 EDSAC (Electronic Delay Storage Automatic Calculator)

1949 IBM CPC (Card-Programmed electronic Calculator)

1952 IBM 701

1952 MANIAC I (Mathematical ANalyzer, Integrator And Computer)

1961 MANIAC III (Mathematical ANalyzer, Integrator And Computer)

I.4. POPULARIZATION

The classification explains why it was not until the article by Metropolis and Ulam (1949), where the name Monte Carlo was first announced publicly that no reference was made to the method. Only limited copies of Ulam's publications (1949) and Householder (1951) were distributed. Subsequently, other publications that had been classified before also appeared (E. Fermi, R. Richtmyer, 1948).

A detailed bibliography of publications related to the Monte Carlo method for the period 1949 to 1963, including in USSR, was published by Kraft and Wensrich (1964). To date N. H. F. Beebe (2017) of the University of Utah maintains a detailed bibliography that updates annually.

In 1949 to 1954, two symposia were organized in relation to the Monte Carlo method. The first one took place at the University of California, Los Angeles (from June 29 to July 1, 1949) (Institute for Numerical Analysis, 1951). There Ulam delivered his paper entitled "Problems in Probability and Combinatorial Analysis" in which he mentioned for the first time the term "Monte Carlo" according to Hurd. Ulam's paper, however, was not published as well as some other papers from the symposium because they treated classified information. The second symposium was held at the University of Florida, Gainesville (March 16-17, 1954) (H. A. Meyer, 1956).

The Monte Carlo method was declassified on July 31, 1954, which provided huge opportunities for its application in practice (F. Harlow, N. Metropolis, 1983).

After the UN Geneva Conference on peaceful uses of atomic energy (1955) (Figure I-15) the first publications related to the method appeared in the Soviet Union (V. Chavchavadze, 1955; N. P. Buslenko, Y. A. Shreider, G. J. Tee, 1966; B. Novozhilov, 1966).

In 1976, Ulam published his autobiography *"Adventures of a Mathematician"*. Three years after his death the Los Alamos

Laboratory (1987) published a special issue of its journal dedicated to Ulam.

Figure I-15. "The Big Four", Geneva, 1955: [from left to right] Nikolai Bulganin (Premier of the Soviet Union), Dwight Eisenhower (President of the USA), Edgar Faure (Prime Minister of France) and Sir Anthony Eden (Prime Minister of Britain)

The Google Books search by key word "Monte Carlo Methods" returns more than 200 thousand results, and, if the keyword is written in Cyrillic, more than 90 thousand results (July 2016)! However, It should be noted that here it is all about the key words mentioned in the contents of the source as they are not necessarily dedicated only to the method.

In 1995, Walter De Gruyter Publishing House started publishing a scientific journal dedicated to the method: "*Monte Carlo Methods and Applications*".

1934 Enrico Fermi experimented with a similar method but published nothing on the subject

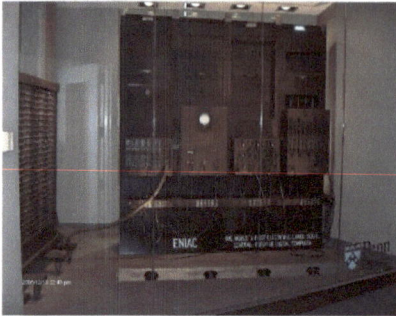

1945 ENIAC was built by J. Mauchly and J. Eckert

1946 S. Ulam describes his idea to von Neumann

1946 E. Fermi built FERMIAC

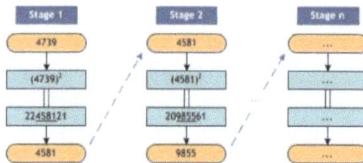

1947 J. von Neumann created the first random number generator

1950 The method was used in Los Alamos during the development of hydrogen bomb

1959 The Monte Carlo method was declassified

1942 President Roosevelt signing the order for launching the Manhattan Project after Einstein–Szilárd letter

1947 The first report related to the Monte Carlo method by Ulam, von Neumann and Richtmyer, which was classified

1948 The first symposium related to the Monte Carlo method; Ulam first mentions the name "Monte Carlo"

JOURNAL OF THE AMERICAN
STATISTICAL ASSOCIATION

Number 247   SEPTEMBER 1949   Volume 44

THE MONTE CARLO METHOD

NICHOLAS METROPOLIS AND S. ULAM
Los Alamos Laboratory

We shall present here the motivation and a general description of a method dealing with a class of problems in mathematical physics. The method is, essentially, a statistical approach to the study of differential equations, or more generally, of integro-differential equations that occur in various branches of the natural sciences.

1949 The first official publication on the method. The name "Monte Carlo" announced

REFERENCES

[1] Abella, A. (2008). Soldiers of Reason: The Rand Corporation and the Rise of the American Empire, Harcourt, Incorporated.

[2] Ackland, L. (1990). The Bulletin of the Atomic Scientists. Chicago, IL, Educational Foundation for Nuclear Science.

[3] Anderson, H. (1986a). "Metropolis, Monte Carlo, and the MANIAC." Los Alamos Science: 96-108.

[4] Anderson, H. (1986b). "Scientific Uses of the MANIAC." Journal of Statistical Physics 43(5/6): 731-748.

[5] Atanasoff, J. V. (1984). "Advent of electronic digital computing." IEEE Ann. Hist. Comput. 6(3): 229-282.

[6] Beebe, N. H. F. (2017). A Complete Bibliography of Publications in Monte Carlo Methods and Applications. Salt Lake City, UT, University of Utah: 119.

[7] Benov, D. (2016). "The Manhattan Project, the first electronic computer and the Monte Carlo method." Monte Carlo Methods and Applications 22(1): 73-79.

[8] Burks, A. (1981). The ENIAC: First general-purpose electronic computer. IEEE Annals of the History of Computing.

[9] Buslenko, N. P., Y. A. Shreider, G. J. Tee (1966). The Monte Carlo Method: The Method of Statistical Trials, Pergamon Press.

[10] Ceruzzi, P. E. (2003). A History of Modern Computing, MIT Press.

[11] Chavchavadze, V. (1955). "Method of random testing: (Monte Carlo method)." Trud 3: 105-121 [in Russian].

[12] Cragon, H. G. (2000). Computer Architecture and Implementation, Cambridge University Press.

[13] De Mol, L. (2009). "Doing Mathematics on the ENIAC. Von Neumann's and Lehmer's different visions.".

[14] Eckhardt, R. (1987). "STAN ULAM, JOHN VON NEUMANN, and the MONTE CARLO METHOD." Los Alamos Science Special Issue: 131-143.

[15] Fermi, E., R. Richtmyer (1948). Note on Census-Taking in Monte-Carlo Calculations. Los Alamos, NM, Los Alamos National Laboratory.

[16] Ford, K. W. (2015). Building the H Bomb: A Personal History, World Scientific Publishing Company Pte Limited.

[17] Gorelik, G. (2011). "The Riddle of the Third Idea: How Did the Soviets Build a Thermonuclear Bomb So Suspiciously Fast?" Scientific American.

[18] Groueff, S. (1967). Manhattan Project: The Untold Story of the Making of the Atomic Bomb. Boston, MA, Little Brown & Co.

[19] Groves, L. (1962). Now It Can Be Told: The Story of the Manhattan Project. New York, Harper.

[20] Harlow, F., N. Metropolis (1983). "Computing&Computers. Weapons Simulation Leads to the Computer Era." Winter/Spring 1983 LOS ALAMOS SCIENCE: 132-141.

[21] Holt, J. (2012). How the Computers Exploded. The New York Review of Books. New York.

[22] Householder, A. (1951). Monte Carlo method. Symposium, Los Angeles, CA.

[23] Hurd, C. (1985). "A Note on Early Monte Carlo Computations and Scientific Meetings." Annals of the History of Computing 7(2): 141-155.

[24] Hyman, A. (1982). Charles Babbage: Pioneer of the Computer, Princeton University Press.

[25] Institute for Numerical Analysis (1951). Monte Carlo Method: Proceedings of a Symposium Held June 29, 30, and July, 1, 1949, in Los Angeles, California, Under the Sponsorship of the Rand Corporation, and the National Bureau of Standards, with the Cooperation of the Oak Ridge National Laboratory, U.S. Government Printing Office.

[26] Isaacson, W. (2014). The Innovators: How a Group of Hackers, Geniuses, and Geeks Created the Digital Revolution, Simon & Schuster.

[27] King, G. (1951). "Monte-Carlo Method for Solving Diffusion Problems." Industrial & Engineering Chemistry 43(11): 2475-2478.

[28] Kraft, R., C. Wensrich (1964). Monte Carlo methods : a bibliography covering the period 1949 to 1963. Berkeley, CA, University of California, Lawrence Radiation Laboratory.

[29] Leeming, J. (1980). Games and Fun with Playing Cards, Dover Publications.

[30] Lilov, E. (2013). And Japan wanted an atomic bomb. Deutsche Welle [in Bulgarian].

[31] Los Alamos National Laboratory (1987). Los Alamos Science 15(Special): 321.

[32] Metropolis, N. (1987). "The Beginning of Monte Carlo." Los Alamos Science Special Issue: 125-130.

[33] Metropolis, N. (2014). History of Computing in the Twentieth Century, Elsevier Science.

[34] Metropolis, N., J. Howlett, G. C. Rota (1980). A History of Computing in the Twentieth Century: A Collection of Essays, Academic Press.

[35] Metropolis, N., S. Ulam (1949). "The Monte Carlo Method." Journal of the American Statistical Association 44(257): 335-341.

[36] Meyer, H. A. (1956). Symposium on Monte Carlo Methods: Held at the University of Florida, Conducted by the Statistical

Laboratory, Sponsored by Wright Air Development Center of the Air Research and Development Command, March 16 and 17, 1954, John Wiley & Sons, Incorporated.

[37] Novozhilov, B. (1966). Method Monte Carlo. Moscow, Znanie [in Russian].

[38] Petzold, C. (2008). The annotated Turing: a guided tour through Alan Turing's historic paper on computability and the Turing machine, Wiley Pub.

[39] Rajchel, T. (2013). Stanislaw Ulam-Player of a Key Role in the Development of the Atomic and Hydrogen Bombs. The Utica Phoenix. Phoenix, AZ.

[40] Richelson, J. (2007). Spying on the Bomb: American Nuclear Intelligence from Nazi Germany to Iran and North Korea, W. W. Norton.

[41] Richtmyer, R., J. von Neumann, S. Ulam (1947). Statistical Methods in Neutron Diffusion. Los Alamos, Los Alamos National Laboratory: 22.

[42] Schließ, G. (2015). Wenn Hölle hinab auf die Erde. Deutsche Welle.

[43] Turing, A. (1936). "On Computable Numbers, with an Application to the Entscheidungsproblem." Proceedings of the London Mathematical Society Series 2(42): 230–265.

[44] Ulam, S. (1949). On the Monte Carlo Method. Symposium on Large Scale Digital Calculating Machines, Cambridge, Mass., Harvard University Press.

[45] Ulam, S. (1991). Adventures of a Mathematician, University of California Press.

[46] Vardi, N. (2012). America's Richest Counties. Forbes. Jersey City, NJ, Forbes Media.

[47] von Neumann, J. (1945). First Draft of a Report on the EDVAC, Moore School of Electrical Engineering, University of Pennsylvenia: 355.

[48] Walter De Gruyter (2016). "Monte Carlo Methods and Applications." from http://www.degruyter.com/view/j/mcma.

SECTION II.
ESSENCE

M. Mazhdrakov, D. Benov, G. Trapov, St. Topalov,
M. Dimov, D. Benova

L et us examine some hypothetical situations. A real estate broker sells a plot of land having an area of 1,000 square meters. The buyer has hired an expert who measures the area and obtains a result of 995 square meters. The market price per 1 square meter is €100, i.e. expressed in money the difference is €500 and thus the broker contests the accuracy of the measurement. Is it possible to establish what the error is?

You have $10^n, n \gg 1$, robotized devices with a keyboard and memory. Is it possible that after entering a certain number of symbols those devices can generate any of Ivan Vazov's poems?

And is it possible that a group of young people who are taking photos by their smart phones can assemble an Oscar-winning movie?

A hopper is full of rock pieces of different dimensions numbered by $i = 1,2,3,...,n$. We know that the pieces are isometric and have a volume v_i. If piece No. 101 with dimensions d_{101} is extracted first, which and how many pieces will follow it, and them?

A gully with length S and vertical section P runs from San Andreas lake to city N. The catchment area is L km^2, and the dam's volume is V m^3. The drainage ratio is from 0.2 to 0.9. The expected quantity of rainfall is from 30 to 110 mm/m^2. Is there a danger of flooding?

m components must be mixed so that the resulting mix has certain property – p_0. To that end researchers use the so-called varying of ingredient one by one and find the combination in which the mix has the properties sought. But, they know at that, that the method does not take into account the joint action of ingredients. How can they help themselves?

Under a design solution the stability ratio of the slopes of a large-scale open-pit mine is 1.15. The boundary value of the ratio is 1.10. To what extent is the established ratio a function of the probable values of the physico-mechanical indicators of the massif?

What are the common characteristics in the hypothetical situations stated, and in countless others? The first one is that there is no unequivocal answer for them, and the second is that in order to get an answer we have to take the probabilistic nature of the phenomena into account.

The solution is in the Monte Carlo method

The effective application of the Monte Carlo method requires the following.

1. To create a mathematical model of the object, process or phenomenon under investigation that is appropriate to the calculations.

2. To realize sufficient number of calculations of the result.

3. To meet the condition of independence of runs.

It took resourcefulness and diligence to overcome such difficulties at the time when the method was created as evident from SECTION I.

The Monte Carlo method may be used for solving tasks of a variety of fields such as finding extreme values of the target function in optimization procedures, message encoding, and other suchlike. However, there is no mention of its successful application in the so-called "games of chance". Despite that there are examples of its application in forecasting the results of sporting events (J. Lahvicka, 2012; A. R. Hassan, J. C. Jimenez, 2014), used by the bookmakers.

The practical application of the Monte Carlo method and the certainty of the results obtained require that certain philosophical concept for the probabilistic nature of our surrounding world be adopted.

In this regard the dispute between Bohr and Einstein which became the focus of lively discussions at the fifth congress of physicists in 1927 organized by Ernest Solvay is quite telling.

What did they disagree about?

Figure II-1. Albert Einstein and Niels Bohr – friends and opponents

Einstein insisted that the principles of determinism of classical physics and the interpretation of the results as "detached observer". According to Einstein there are some hidden variables with the help of which results can be forecasted and quantum physics can be statistically defined.

Bohr insisted on the statistical nature of quantum phenomena, which is non-determined in principle, and that the effect of measurements is non-eliminable.

99 *Einstein — God doesn't play dice*

Bohr — Stop telling God what to do with his dice

Einstein — Do you really believe that the moon isn't there when nobody looks?

Physicists were divided into two camps. Planck and Schrödinger took Einstein's side, and Heisenberg, Born and Dirac took Bohr's.

The essence of the dispute is: Whether he world is in fact governed by indeterminacy or simply we do not know some properties of microparticles and if we can measure them we will be able to foresee their behavior in any specific situation (M. Kumar, 2011).

Today the popular question is: *"Can a butterfly flying over Beijing cause a hurricane in the Caribbean?"* The accurate answer is: *"Yes, it can but this is quite unlikely!"*

II.1. STOCHASTIC MODELING

The Monte Carlo method is defined as a *"Method of stochastic modeling"* or a *"Method of statistical modeling"*. By definition these two characteristics are synonymous as they refer to processing of random values.

Random value means a variable, which depending on the circumstances gets, with certain probability, some random (different, unequivocally unforeseeable) value. The assumption that there exist random values is a matter of worldview but this philosophical problem is beyond their mathematical treatment.

There is certain difference between the approaches in random value modeling in "classical" mathematical statistics and in the Monte Carlo method. Mathematical statistics reduces modeling to finding the coefficients of given functions: polynomials of different degree, exponential, trigonometric, etc., for which a physical justification exists. For example, several proven correlations are observed in prospecting and extraction of ores and minerals: between coal volumetric density and ash content, between the thickness of the vein of ore and the content of useful component, between the contents of some metals, etc. In all cases the regression function is of the same type

$$y = ax + b, \qquad\qquad (II.1)$$

where a and b are coefficients calculated for each particular case.

In contrast to that the Monte Carlo method is based on methods reflecting the specific peculiarities of the modeled object, or, process, respectively, including ones for which the application of the apparatus of mathematical analysis is difficult or practically impossible.

The word "stochastic" comes from the Greek word "στόχος", meaning perspicacity, skill to foresee, and "statistics" comes from the Latin word "status", i.e. a "state". This is why we qualify the Monte Carlo method as a method for stochastic modeling; by the way, this was made as early as in the first publications where it was pointed out that the Monte Carlo method combines statistics and mathematical analysis.

It is accepted that the idea of combining the two approaches was first voiced by the French mathematician and naturalist Georges-Louis Leclerc, Comte de Buffon.

The characteristic of "simulation modeling" used is associated with two Latin words of similar spelling but different meaning: "simulare" meaning "to pretend" and "simula" meaning "together", "simultaneously".

II.2. LAW OF LARGE NUMBERS

According to the law of large numbers if the number of realizations of a random value is great the result obtained does not depend on the case due to the joint action of various random factors. Then an approximation of the frequency of appearance of the random event with the probability of its appearance is observed. This objective law was first observed in the games of chance in the Middle Ages.

The first mathematical proof of the law of large numbers is Jacob Bernoulli's Theorem: In the series of n independent realizations in which the probability of the occurrence of the random event $A - p$, $(0 < p < 1)$, is the same, and where $n \to \infty$, then the inequality is valid

$$p\left\{\left|\frac{\mu_n}{n} - p\right| > \varepsilon\right\} \to 0, \tag{II.2}$$

where $\varepsilon > 0$;

$\frac{\mu_n}{n}$ is the frequency of realization of the event A for the first n realizations.

Jacob Bernoulli's Theorem was extended by S. Poisson for the case where the probability of realization of the event A depends on the trial number.

The method of stochastic modeling realizes in practice the law of large numbers (A. Obretenov, 1978; I. M. Sobol, 1994). According to this law, if $\xi_1, \xi_2, \xi_3, ..., \xi_n$ are independent random values with the same mathematical expectation of $E\xi_i = a$, if the values of n are large, then the approximate equality is fulfilled

$$\frac{\xi_1 + \xi_2 + \xi_3 + \cdots + \xi_n}{n} \approx a, \tag{II.3}$$

where a is the unknown characteristic of a phenomenon or a process.

As a rule the immediate measurement of a is difficult or impossible. However, it is possible to reproduce a random experiment in which the random value of ξ can be defined by mathematical expectation $E\xi = a$. Then, by means of independent serial realizations of ξ the observations $\xi_1, \xi_2, \xi_3, \ldots, \xi_N$. can be simulated.

In practice, although the simulated observations are a finite number N, on the basis of the law of large numbers it can be assumed that the number

$$\bar{\xi} = \frac{\xi_1 + \xi_2 + \xi_3 + \cdots + \xi_N}{N} \qquad \text{(II.4)}$$

is a sufficiently precise evaluation of the unknown characteristic a.

Jacob Bernoulli (1654-1705) Swiss mathematician of Dutch origin

Georges-Louis Leclerc, Comte de Buffon (1707-1788) French mathematician, naturalist and cosmologist

Siméon Poisson (1781-1840) French mathematician, geometrician and physicist

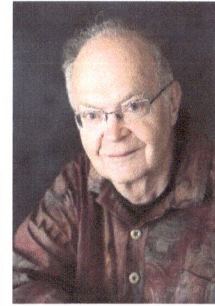

Donald Knuth (1938) American theoretician-informant, father of algorithm analysis

In the past, due to the absence of quick-acting computing machines to carry out a sufficient number of runs, different computing schemes were applied and the calculations were divided into groups and/or separate performers.

The creation of computing means has eliminated the said obstacles to a great extent and has enabled the wide application of the Monte Carlo method in virtually all fields of human activity.

II.3. RANDOM NUMBERS

The requirement for independence of data and their distribution along the entire possible spectrum is realized by using the so-called "random numbers". Random numbers are a sequence of numbers where each number is independent of its predecessor and the probability of its appearance corresponds to a certain, most often uniform, distribution, in the interval [0;1), whence it is not difficult to proceed to some theoretical or to an empirical distribution.

Methods of random number generation have existed for a long time, for example: rolling of dice, coin flipping, shuffling of playing cards, etc. It has been intuitively assumed that random numbers are generated when spinning the roulette wheel, whence the name of the method. Subsequently, it turned out that this hypothesis is not true and random number tables were used.

The first random number table was made by L. H. C. Tippett (1927). He took the census register. After him R. Fisher and F. Yates used logarithm tables to make their random number table. In 1939, M. Kendall and B. Babington Smith published a table of 100,000 random digits produced by an operator-controlled machine (M. Kendall, B. Babington Smith, 1939).

TABLE OF RANDOM DIGITS   1

00000	10097	32533	76520	13586	34673	54876	80959	09117	39292	74945
00001	37542	04805	64894	74296	24805	24037	20636	10402	00822	91665
00002	08422	68953	19645	09303	23209	02560	15953	34764	35080	33606
00003	99019	02529	09376	70715	38311	31165	88676	74397	04436	27659
00004	12807	99970	80157	36147	64032	36653	98951	16877	12171	76833
00005	66065	74717	34072	76850	36697	36170	65813	39885	11199	29170
00006	31060	10805	45571	82406	35303	42614	86799	07439	23403	09732
00007	85269	77602	02051	65692	68665	74818	73053	85247	18623	88579
00008	63573	32135	05325	47048	90553	57548	28468	28709	83491	25624
00009	73796	45753	03529	64778	35808	34282	60935	20344	35273	88435
00010	98520	17767	14905	68607	22109	40558	60970	93433	50500	73998
00011	11805	05431	39808	27732	50725	68248	29405	24201	52775	67851
00012	83452	99634	06288	98083	13746	70078	18475	40610	68711	77817
00013	88685	40200	86507	58401	36766	67951	90364	76493	29609	11062
00014	99594	67348	87517	64969	91826	08928	93785	61368	23478	34113

Figure II-2. Excerpts from the Table of Random Digits of The RAND Corporation (1955) [13]

---

[13] Third party copyright. © RAND Corporation, Santa Monica, CA, www.rand.org. the sample is published with explicit written permission.
The full table of random digits of the RAND Corporation is available at
http://www.rand.org/pubs/monograph_reports/MR1418.html.

In mid 1940s the associates of the RAND Corporation developed an impressive random number table for the use with the Monte Carlo method (Figure II-2). Using a hardware random number generator - an electronic simulation of a roulette wheel attached to a computer – they produced one million random digits and a table of 100,000 normal deviates (The RAND Corporation, 1955). He results were carefully tested before being added to the table. This set of random numbers was exceptionally important because such a large and carefully prepared random number table had never before existed. Another important advantage of the table was that it was made on IBM punched cards and was machine-readable.

When using random number tables, however, the runs were extremely slow. For that reason Von Neumann developed an algorithm for calculation of pseudo-random numbers – The Middle-Square Method (J. von Neumann, 1951). Von Neumann's algorithm is

$$R_{n+1} = \mathrm{mid}(R_n^2, m), \qquad\qquad (II.5)$$

where R_{n+1} is the new random number;

R_n – is the preceding random number;

$\mathrm{mid}(x, m)$ – the middle m digits of the number x.

The algorithm consists in the following steps.

1. One starts with a m-digit starting value.

2. The starting value is squared to produce a $2m$-digit value; if necessary more zeros are added.

3. The middle m-digits are the next random number.

4. The computation scheme is repeated as many times as necessary.

We will examine an example of realization of the method (II.5):

$$1307 \rightarrow 1307^2 = 01708249 \rightarrow 7082 \rightarrow 7082^2 = 50154724$$
$$\rightarrow 1547 \rightarrow 1547^2 = 02393209 \rightarrow 3932 \rightarrow \cdots$$

Another example of realization is given in Figure II-3.

The approach is close to the solution based on the lists of residents of populated areas of 10,000 – 99,999 residents where random numbers are 000 to 999.

There are also starting numbers where the method does not take effect

$$7600 \rightarrow 7600^2 = 57760000 \rightarrow 7600 \rightarrow 7600^2 = 57760000 \rightarrow \cdots$$

Currently, the Monte Carlo method is realized by using random number generators.

Table II-1. Random numbers table (N. Smirnov, I. Dunin-Barkovskii, 1962)

1393	6270	4228	6069	9407	1865	8549	3217	2351	8410
9108	2330	2157	7416	0398	6173	1703	8132	9065	6717
7891	3590	2502	5945	3402	0491	4328	2365	6175	7695
9085	6307	6910	9174	1753	1797	9229	3422	9861	8357
2638	2908	6368	0398	5495	3283	0031	5955	6544	3883
1313	8338	0623	8600	4950	5414	7131	0134	7241	0651
3897	4202	3814	3505	1599	1649	2784	1994	5775	1406
4380	9543	1646	2850	8415	9120	8062	2421	6161	4634
1618	6309	7909	0874	0401	4301	4517	9197	3350	0434
4858	4676	7363	9141	6133	0549	1972	3461	7116	1496
5354	9142	0847	5393	5416	6505	7156	5634	9703	6221
0905	6986	9396	3975	9255	0537	2479	4589	0562	5345
1420	0470	8679	2328	3939	1292	0406	5428	3789	2882
3218	9080	6604	1813	8209	7039	2086	3369	4437	3798
9697	8431	4387	0622	6893	8788	2320	9358	5904	9539
0912	4964	0502	9683	4636	2861	2876	1273	7870	2030
4636	7072	4868	0601	3894	7182	8417	2367	7032	1003
2515	4734	9878	6761	5636	2949	3979	8650	3430	0635
5964	0412	5012	2369	6461	0678	3693	2928	3740	8047
7848	1523	7904	1521	1455	7089	8094	9872	0898	7174
5192	2571	3643	0707	3434	6818	5729	8614	4298	4129
8438	8325	9886	1805	0226	2310	3675	5058	2515	2388
8166	6349	0319	5436	6838	2460	6433	0644	7428	8556
9158	8263	6504	2562	1160	1526	1816	9690	1215	9590
6061	3525	4048	0382	4224	7148	8259	6526	5340	4064

Random number generator is a computing or physical device that generates a sequence of numbers (or symbols) lacking any functional relation, i.e. they are random.

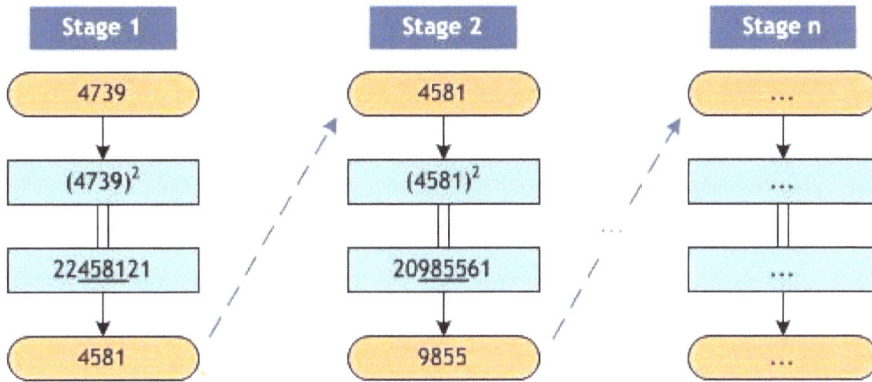

Figure II-3. Model realization of the Middle-Square Method with a starting value of 4739 (University of Notre Dame, 2006)

Listing II-1. Model code for The Middle-Square Method (language: C)

```
1.   #include <stdio.h>
2.   #include <math.h>
3.   #include <stdlib.h>
4.
5.   /*Von Neumann Algorithm*/
6.   /*Programmers: Gursharn, Smruti, Shwetab, Sagnika, Kaustav*/
7.   /*Date: Aug 14, 2010*/
8.
9.   unsigned long int randm(int n);
10.  unsigned long int von(unsigned long int x, int n);
11.
12.  int main(void)
13.  {
14.    unsigned long int x, s;
15.    int n;
16.    printf("Enter the number of digits in the seed value ");
17.    scanf("%d",&n);
18.    if (n >= 10){
19.      printf("TOO LARGE!!");
20.      exit(0);
21.    }
22.    x = randm(n);
23.    printf("\nRandom Number generated: %lu\n",von(x,n));
24.    return 0;
25.  }
26.
```

```
27.
28. /*Generating Random Number of desired digit*/
29.
30. unsigned long int randm(int n)
31. {
32.    double x;
33.    unsigned long int y;
34.    srand(getpid());
35.    x = rand()/(double)RAND_MAX;
36.    y =(int) (x * pow(10.0,(double) n));
37.    return y;
38. }
39.
40.
41. /*Calculating Random Number By Von Neumann method*/
42.
43. unsigned long int von(unsigned long int x, int n)
44. {
45.    unsigned long int y;
46.    int k;
47.    k = n/2;
48.    y = (int)((x/pow(10.0, k * 1.0)) * x);
49.    y = y % (int) (pow(10.0, n * 1.0));
50.    return y;
51. }
52.
53.
54. Code for generating sequence of random number:
55.
56. #include <stdio.h>
57. #include <math.h>
58. #include <stdlib.h>
59.
60. /*Von Neumann Algorithm To Generate Several Random Numbers*/
61. /*Programmers: Gursharn, Smruti, Shwetab, Sagnika, Kaustav*/
62. /*Date: Aug 14, 2010*/
63.
64. unsigned long long int randm(int n);
65. unsigned long long int von(unsigned long long int x, int n);
66.
67. int main(void)
68. {
69.    unsigned long long int x, s;
70.    int n, i, r;
71.
72.    printf("Enter the number of digits in the seed value ");
73.    scanf("%d",&n);
74.
75.    printf("Enter the total random numbers to be generated ");
76.    scanf("%d",&r);
77.
78.    if (n >=12){
```

```
79.        printf("TOO LARGE!!");
80.        exit(0);
81.    }
82.
83.    x = randm(n);
84.    for(i = 0; i < r; i++){
85.        s = von(x,n);
86.        x = s;
87.    printf("\nRandom Number generated: %lld\n",s);
88.    }
89.    return 0;
90. }
91.
92.
93. /*Generating Random Number of desired digit*/
94.
95. unsigned long long int randm(int n)
96. {
97.    double x;
98.    unsigned long long int y;
99.    srand(getpid());
100.      x = rand()/(double)RAND_MAX;
101.      y =(unsigned long long int) (x * pow(10.0, n*1.0));
102.      return y;
103.   }
104.
105.
106.   /*Calculating Random Number By Von Neumann method*/
107.
108.   unsigned long long int von(unsigned long long int x, int n)
109.   {
110.      unsigned long long int y;
111.      int k;
112.      k = n/2;
113.      y =(unsigned long long int)((x/pow(10.0, k * 1.0)) * x) % (u
    nsigned  long long int) (pow(10.0, n * 1.0));
114.      return y;
115.   }
```

Software random number generators (RNG) are algorithms generating a long sequence of numbers meeting the statistical tests for "randomness". The numbers in such a sequence, however, start to repeat themselves sooner or later.

As we have said, program generators do not produce random numbers strictly speaking because:

– the same sequences are produced by the same starting number;

— certain periodicity can be observed, even if after a considerable number of runs.

For this reason it is accepted to refer to such numbers as "*pseudorandom*" ones (J. Gentle, 2005).

> *Anyone who attempts to generate random numbers by deterministic means is, of course, living in a state of sin.*
>
> — *John von Neumann*

> *Many random number generators in use today are not very good. There is a tendency for people to avoid learning anything about such subroutines; quite often we find that some old method that is comparatively unsatisfactory has blindly been passed down from one programmer to another, and today's users have no understanding of its limitations.*
>
> — *Donald Knuth (1968)*

Because of these peculiarities the software random number generators are not suitable for modeling of random procedures where:

— what matters is the unit value of the quantity being examined and not the frequency of its occurrence;

— no systematic (periodic) change of the quantity being modeled should be allowed in multiple starts from the beginning.

Let us examine the following distribution problem. N children should get N different presents from Santa Claus. The conditions are:

— each child should get 1 present;

— the distribution of presents should be random, i.e. it should not depend on the characteristics of the recipient (a girl or a boy, older or younger, etc.).

Then, if we adopt the Monte Carlo model we are seeking to obtain two ordered (but not consecutive) sets of N numbers each: one for the children, another for the presents. The problem is solved where the above condition is fulfilled. The result obtained is on of the type:

Child		1	2	3	4	5	6	7	8	...
Present		33	21	13	15	2	7	24	12	...

So what's the trick? It is in the fact that after every run the same sequences will be generated.

What should "our girl" do to win a talking Barbie doll, which is present No. 13?

1. The program is performed and it is established that after the first successful run present No. 13 is at the 3rd place.

2. The run is discontinued in emergency because of a volcano eruption in Krasno Selo.

3. Our girl gets present No. 3.

4. Operation No. 1 is repeated and Santa Claus gives away presents.

5. Due to a logistical error our girl gets a Cristiano Ronaldo T-shirt.

This, of course, is a harmless example but a similar realization was made in practice by Eddie Raymond Tipton who used his official position to obtain personal gain (AP, 2015). The information security employee at the US Lottery won millions of dollars for several years by slightly changing the lottery's random number generator, which enabled him to "guess" the winning numbers several times a year.

Tipton was arrested in 2015 and it became known how he had managed to get rich. As information security director of Multi-State Lottery Association he had access to the premises where the random number generator is situated. He changed the system software by adding his own code. This happened after the generator had been checked by an independent external company. By means of the added code the generator does not count random numbers but generates them by "Tipton's algorithm". In this way three times a year he could guess the winning numbers. He used this opportunity as many as six times from November 23rd, 2005 to December 29th, 2011. The specialists studied his combinations and found the pattern behind them. Had Tipton made use of this opportunity only one or two times there would be no way to find a pattern and hardly could anyone suspect a thing. External libraries creating a predictable combination three times a year on one of the two defined days of the week were inserted into the computer.

Figure II-4. The draws for the UEFA Champions League (2013)

Of course, to carry out the rigging as aforesaid the so-called "logic bombs" must be inserted into the software to realize the peculiarity of program random number generators. In this case the solution is to use a physical random number generator, e.g. a bag with sheets of paper, a cup of marbles, etc. similarly to the rigged draws in FIFA and WEFA championships (Figure II-4).[16] (The Guardian, 2016).

The problem with evaluating the "randomness" of random numbers generated by different methods has accompanied most known solutions since the time of Tippett (1927).

The first algorithm for "testing" random numbers for statistical randomness was developed by Kendall and Babington-Smith (1938) and is based on looking for certain types of probabilistic expectations in a given sequence.

The tests for frequency of occurrence are based on the known criteria for evaluation of the proximity of empirical distributions and normal distribution. The most famous of these tests is the χ^2 criterion and the Kolmogorov–Smirnov criterion (W. Eadie, D. Drijard, F. James et al., 1971).

The frequency of occurrence of the same random number is evaluated by the length of the interval determined by the number of digits between recurrences (the so-called "gap test").

The tests for evaluation of series of generated numbers check to what extent the difference between two consecutive random numbers is uniformly distributed within the expected interval. Other versions of the tests are known, too. For example if one follows the frequency of occurrence of a series of 5 consecutive numbers, this is the so-called "poker test".

If it's hard to distinguish a generator's output from truly random sequences the generator is assessed as a high quality generator; otherwise, it is a low quality generator (M. Malone, 2015).

There is no universally valid formula for calculating sequences of random numbers and thus the generators work by different algorithms, most often by recurrent division and differentiation of the fractional part of the quotient obtained (G. E.

---

[16] However, some marbles are said to have been heated or cooled in order to rig the draw.

Forsythe, M. A. Malcolm, C. B. Moler, 1977; V. Ivanova, 1984; G. Trapov, 2012).

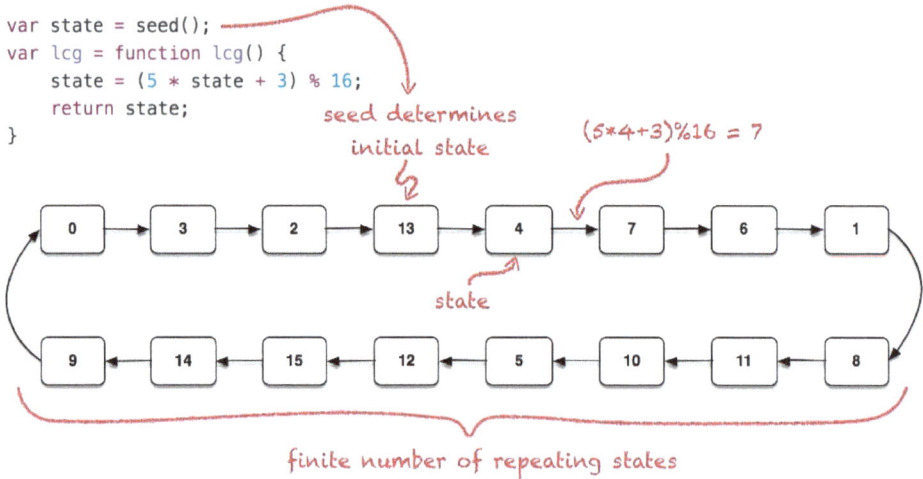

```
var state = seed();
var lcg = function lcg() {
    state = (5 * state + 3) % 16;
    return state;
}
```

seed determines
initial state

(5*4+3)%16 = 7

state

finite number of repeating states

Figure II-5. *Sample algorithm for a random number generator (M. Malone, 2015)*

Linear congruential method (LCG)

One of the most famous algorithms for random number generation is the linear congruential method (LCG), created in 1951 by Derrick Lehmer (D. Knuth, 1968; K. Entacher, 1997)

$$X_{n+1} = (aX_n + b) \bmod m, \qquad (II.6)$$

where a and b are coefficients;

m is the maximum number of generated numbers.

In the formula (II.6) mod means the remainder of integer division. For example, 11 mod 4 is found by the following manner –

$$11 : 4 = 2 \ (rest \ 3),$$

i.e.

$$11 \bmod 4 = 3.$$

From that it follows that whenever a number recurs in a sequence of random numbers generated all numbers generated thereafter will also recur.

The selection of coefficients a and m is of particular importance. In case of correct selection two aims are attained.

1. There is no correlation between the consecutive numbers.

2. The period of recurrence is as long as possible.

It is recommended that the m number be as big as possible as the sequence of random numbers is in the $[0; m-1]$ interval. If we want to obtain uniformly distributed numbers from 0 to 1 we should divide the m numbers obtained as per the formula (II.6). The first number can be random. Most often it is selected by the computer upon the initialization of the generator. The largest number the processor can operate with also matters.

The linear congruential method is used by virtually all contemporary programming languages. As an example we will examine the realizations in BSD libc and Microsoft C Runtime.

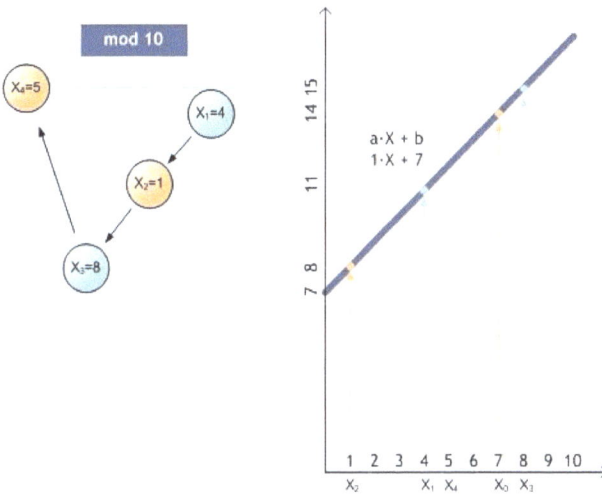

Figure II-6. Model realization of the linear congruential method (University of Notre Dame, 2006)

B BSD libc is used the formula

$$X_{n+1} = (1103515245 \times X_n + 12345) \bmod 2^{31},$$
$$rand_n = X_n \in [0; 2147483647],$$

(II.7)

and in Microsoft C Runtime −

$$X_{n+1} = (214013 \times X_n + 2531011) \bmod 2^{31}.$$
$$rand_n = X_n \div 2^{16} \in [0; 32767].$$
(II.8)

Table II-2 shows the values of a, b and m in the main realizations of the method (J. Cruz, 2006).

A model code of the language C# for generation of n random numbers by using a formula (II.8) is shown in Listing II-2. In the example $n = 10$.

The computation of random numbers in the .NET platform is based on the algorithm of D. Knuth (1968).

Another algorithm for random number generation is Mersenne Twister, created by M. Matsumoto, T. Nishimura (1998). The algorithm is based on the Mersenne prime number (H. Regius, 1536). It is exceptionally fast for computation and was created mainly to be used in the Monte Carlo method.

Several years later, M. Saito, M. Matsumoto (2006) created a variant of the Mersenne Twister algorithm, called SIMD-oriented Fast Mersenne Twister (SFMT).

Table II-2. Values of a, b and m in different variants of the linear congruential method

LCG variant	Constants		
	a	b	m
IBM RANDU	65 539	0	2^{31}
MINSTD	16 807	0	$2^{31}-1$
VAX MTHSRANDOM	69 069	1	2^{32}
BSD rand()	1 103 515 245	12 345	2^{31}
UNIX rand48()	25 214 903 917	11	2^{48}
Cray rand48()	44 485 709 377 909	0	2^{48}

Listing II-2. Model random number generator using the linear congruential method (language: C#)

```csharp
1.  public class LCG
2.  {
3.      private int _state;
4.
5.      public LCG()
6.      {
7.          _state = (int)DateTime.Now.Ticks;
```

```
8.      }
9.
10.     public LCG(int n)
11.     {
12.         _state = n;
13.     }
14.
15.     public int Next()
16.     {
17.         return ((_state = 214013*_state+2531011)&int.MaxValue)>>1
   6;
18.     }
19.
20.     public IEnumerable<int> Seq()
21.     {
22.         while (true)
23.         {
24.             yield return Next();
25.         }
26.     }
27. }
28.
29. class Program
30. {
31.     static void Main()
32.     {
33.         LCG lcg = new LCG(0);
34.         lcg.Seq().Take(10).ToList().ForEach(Console.WriteLine);
35.         Console.ReadKey();
36.     }
37. }
```

Listing II-3. Mersenne Twister generator (language: C#)

```
1.  public class RandomMersenneTwister
2.  {
3.      private const int    N            = 624;
4.      private const int    M            = 397;
5.      private const uint   K            = 0x9908B0DFU;
6.      private const uint   DEFAULT_SEED = 4357;
7.
8.      private ulong []     state        = new ulong[N+1];
9.      private int          next         = 0;
10.     private ulong        seedValue;
11.
12.
13.     public RandomMersenneTwister()
14.     {
15.         SeedMT(DEFAULT_SEED);
16.     }
17.     public RandomMersenneTwister(ulong _seed)
18.     {
```

```
19.            seedValue = _seed;
20.            SeedMT(seedValue);
21.        }
22.
23.        public ulong RandomInt()
24.        {
25.            ulong y;
26.
27.            if((next + 1) > N)
28.                return(ReloadMT());
29.
30.            y  = state[next++];
31.            y ^= (y >> 11);
32.            y ^= (y <<  7) & 0x9D2C5680U;
33.            y ^= (y << 15) & 0xEFC60000U;
34.            return(y ^ (y >> 18));
35.        }
36.
37.        private void SeedMT(ulong _seed)
38.        {
39.            ulong x = (_seed | 1U) & 0xFFFFFFFFU;
40.            int j = N;
41.
42.            for(j = N; j >=0; j--)
43.            {
44.                state[j] = (x*=69069U) & 0xFFFFFFFFU;
45.            }
46.            next = 0;
47.        }
48.
49.        public int RandomRange(int lo, int hi)
50.        {
51.            return (Math.Abs((int)RandomInt() % (hi - lo + 1)) + lo)
52.        }
53. }
```

Listing II-4. Park-Miller-Carta generator (language: LUA)

```
1.   --[[
2.   Copyright (c) 2009 Michael Baczynski, http://www.polygonal.de
3.
4.   Permission is hereby granted, free of charge, to any person
     obtaining
5.   a copy of this software and associated documentation files (the
6.   "Software"), to deal in the Software without restriction, including
7.   without limitation the rights to use, copy, modify, merge, publish,
8.   distribute, sublicense, and/or sell copies of the Software, and to
9.   permit persons to whom the Software is furnished to do so, subject
     to
10.  the following conditions:
11.  The above copyright notice and this permission notice shall be
12.  included in all copies or substantial portions of the Software.
13.  THE SOFTWARE IS PROVIDED "AS IS", WITHOUT WARRANTY OF ANY KIND,
```

```lua
14. EXPRESS OR IMPLIED, INCLUDING BUT NOT LIMITED TO THE WARRANTIES OF
15. MERCHANTABILITY, FITNESS FOR A PARTICULAR PURPOSE AND
16. NONINFRINGEMENT. IN NO EVENT SHALL THE AUTHORS OR COPYRIGHT HOLDERS
    BE
17. LIABLE FOR ANY CLAIM, DAMAGES OR OTHER LIABILITY, WHETHER IN AN
    ACTION
18. OF CONTRACT, TORT OR OTHERWISE, ARISING FROM, OUT OF OR IN
    CONNECTION
19. WITH THE SOFTWARE OR THE USE OR OTHER DEALINGS IN THE SOFTWARE.
20. --]]
21. -- Implement. of the Park Miller (1988) "minimal standard" linear
22. -- congruential pseudo-random number generator.
23. -- @author Michael Baczynski, www.polygonal.de
24. -- @author Thomas R. Koll, www.ananasblau.com
25. -- MIT License
26. local PM_PRNG
27. do
28.   local _parent_0 = nil
29.   local _base_0 = {
30.     prime = math.pow(2, 31) - 1,
31.     nextInt = function(self)
32.       return self:generate()
33.     end,
34.     nextDouble = function(self)
35.       return self:generate() / self.prime
36.     end,
37.     nextIntRange = function(self, min, max)
38.       min = min - 0.4999
39.       max = max + 0.4999
40.       return math.floor(0.5 + min + ((max - min) *
   self:nextDouble()))
41.     end,
42.     nextDoubleRange = function(self, min, max)
43.       return min + ((max - min) * self:nextDouble())
44.     end,
45.     generate = function(self)
46.       self.seed = (self.seed * 16807) % self.prime
47.       return self.seed
48.     end
49.   }
50.   _base_0.__index = _base_0
51.   if _parent_0 then
52.     setmetatable(_base_0, _parent_0.__base)
53.   end
54.   local _class_0 = setmetatable({
55.     __init = function(self, seed)
56.       self.seed = seed or 1
57.     end,
58.     __base = _base_0,
59.     __name = "PM_PRNG",
60.     __parent = _parent_0
61.   }, {
62.     __index = function(cls, name)
63.       local val = rawget(_base_0, name)
64.       if val == nil and _parent_0 then
```

```
65.            return _parent_0[name]
66.          else
67.            return val
68.          end
69.        end,
70.        __call = function(cls, ...)
71.          local _self_0 = setmetatable({}, _base_0)
72.          cls.__init(_self_0, ...)
73.          return _self_0
74.        end
75.      })
76.      _base_0.__class = _class_0
77.      if _parent_0 and _parent_0.__inherited then
78.        _parent_0.__inherited(_parent_0, _class_0)
79.      end
80.      PM_PRNG = _class_0
81.  end
```

Another popular algorithm is Park-Miller algorithm, which is based on the linear congruential method.

The Fibonacci generator is a development of the linear congruential method –

$$X_{n+1} = (X_n + X_{n-1}) \bmod m, \qquad (\text{II.9})$$

where $m \in \text{IN}; m \gg 1;$

$$U_n = \frac{X_n}{m}.$$

II.4. PROBABILITY DISTRIBUTION

In most engineering applications of the Monte Carlo method, output data and calculation results are introduced and presented, respectively, as a probability distribution. This distribution shows the probability $(p_1, p_2, ..., p_n)$ of the occurrence of the individual random event as part of some random process (X), i.e. $p(X = x_i)$. By definition

$$p_i > 0 \text{ and } \sum_{i=1}^{n} p_i = 1.$$

In the examined field the distribution of the probability is obtained directly (upon measurement or observation) or after certain processing by meeting the requirements for independence of observations and the relative constancy of external conditions and methodology of measurement characterizing the random processes.

The results obtained are divided into groups and for each group the probability

$$p_k = \frac{n_k}{\sum_{i=1}^{n} n_i} < 1.$$

The number of intervals must correspond to the number of measurements. A popular formula is

$$N_{gr} = 1 + 3.22 \lg N, \tag{II.10}$$

where N is the number of all observations.

Computational practice shows that if $N_{gr} < 8$ some local changes can be "smoothed" and in case of a great number of $N_{gr} > 20$) the effect of local extremums can be increased, including a reset of some intervals.

The empirical distribution is the relation between the variants of the studied indicator and their frequency of occurrence established in the examination.

Table II-3. Theoretical distributions used often in prospecting and extraction of ores and minerals

Type of theoretical distribution	Scope of application
Normal	- physico-mechanical indicators; - random errors in measurements; - content of useful and harmful components (for deposits with low changeability of indicators); - intensity of movement along transport thoroughfares.
Gamma	- content of useful and harmful components (for deposits with high changeability of indicators).
Exponential	- vectors of transfer and deformation; - indicators of some technological processes.
Δ - Dirak	- output of machines.
Uniform	- rolling a well-balanced die.

In many cases the application of the empirical distribution is associated with certain difficulties. Then the connection between the value of the indicator and the probability of realization is obtained with a class of functions called theoretical distribution. In these distributions the probabilities $p_1, p_2, ..., p_n$ are calculated from a limited number of parameters.

Figure II-7 shows the graphics of some more popular theoretical distributions. The choice of a particular formula depends on the specific properties of the indicator being modeled. For example, in prospecting and extraction of ores and minerals most researchers use the distributions shown in Table II-3 (L. Dimov, 1979).

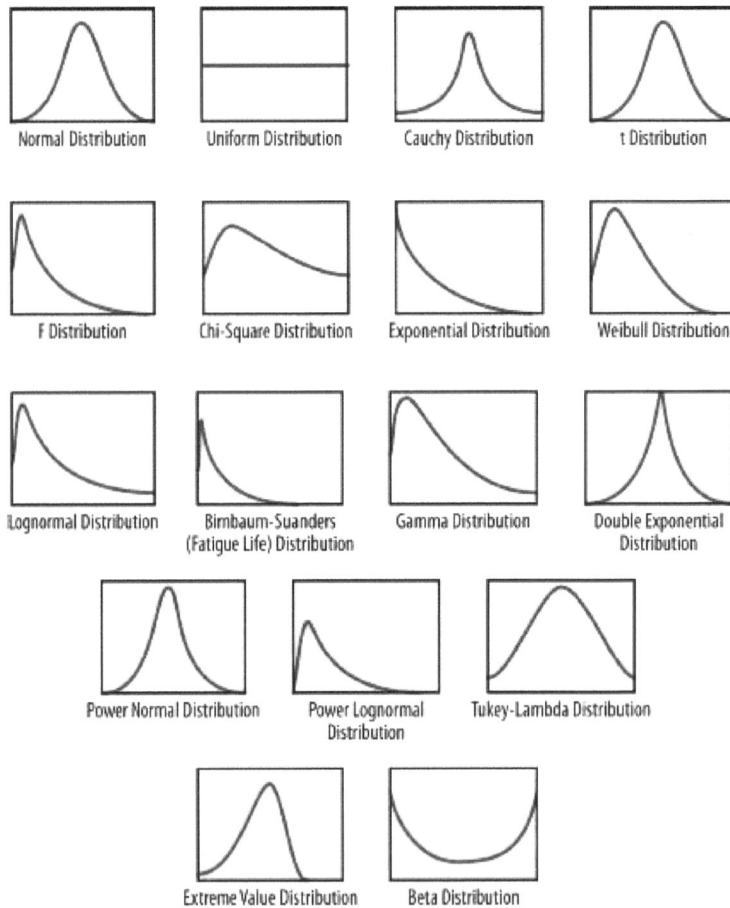

| Normal Distribution | Uniform Distribution | Cauchy Distribution | t Distribution |

| F Distribution | Chi-Square Distribution | Exponential Distribution | Weibull Distribution |

| Lognormal Distribution | Birnbaum-Suanders (Fatigue Life) Distribution | Gamma Distribution | Double Exponential Distribution |

| Power Normal Distribution | Power Lognormal Distribution | Tukey-Lambda Distribution |

| Extreme Value Distribution | Beta Distribution |

Figure II-7. Graphics of some more popular theoretical distributions (R. Schutt, C. O'Neil, 2014)

In the general case the algorithm from Figure II-8 can be used to determine the type of theoretical distribution.

An advantage of the theoretical distribution is that it is set with a limited number of parameters; for example, the uniform distribution – with the number of classes, normal one – with x_{mid} and dispersion – σ_x, etc. A disadvantage is the smoothing of results, especially in unimodal distributions.

Where a random number generator is applied in the interval $[0; 1)$ it is suitable to use the so-called cumulative distribution (Figure II-9), where the frequencies are accumulated

$$c_1 = p_1$$
$$c_2 = c_1 + p_2 = p_1 + p_2$$
$$c_3 = c_2 + p_3 = p_1 + p_2 + p_3$$
$$...$$
$$c_n = 1.$$

As a result of the modeling under the Monte Carlo method one obtains distributions which are close to the ones of the output values. To obtain uniform distributions at entrance and exit is not, however, proof that entry distribution has been appropriately selected because of the effect of self-fulfilling prophecy. Besides, according to Joan Borysenko "Perhaps the hardest lesson to learn is not to be attached to the results of your actions"[14].

As a rule, the results of stochastic modeling under the Monte Carlo method are presented as a distribution of the relative frequency of occurrence of an event. For the purpose of comparability the results are grouped in the range $[X_{min}; X_{max}]$ at equal intervals. According to Sturges' formula (II.10), these groups are 20 in case of sufficiently large number of data; the same is also confirmed by our practice with the application of the method.

---

14 Joan Borissenko (1945) – psychologist and sociologist

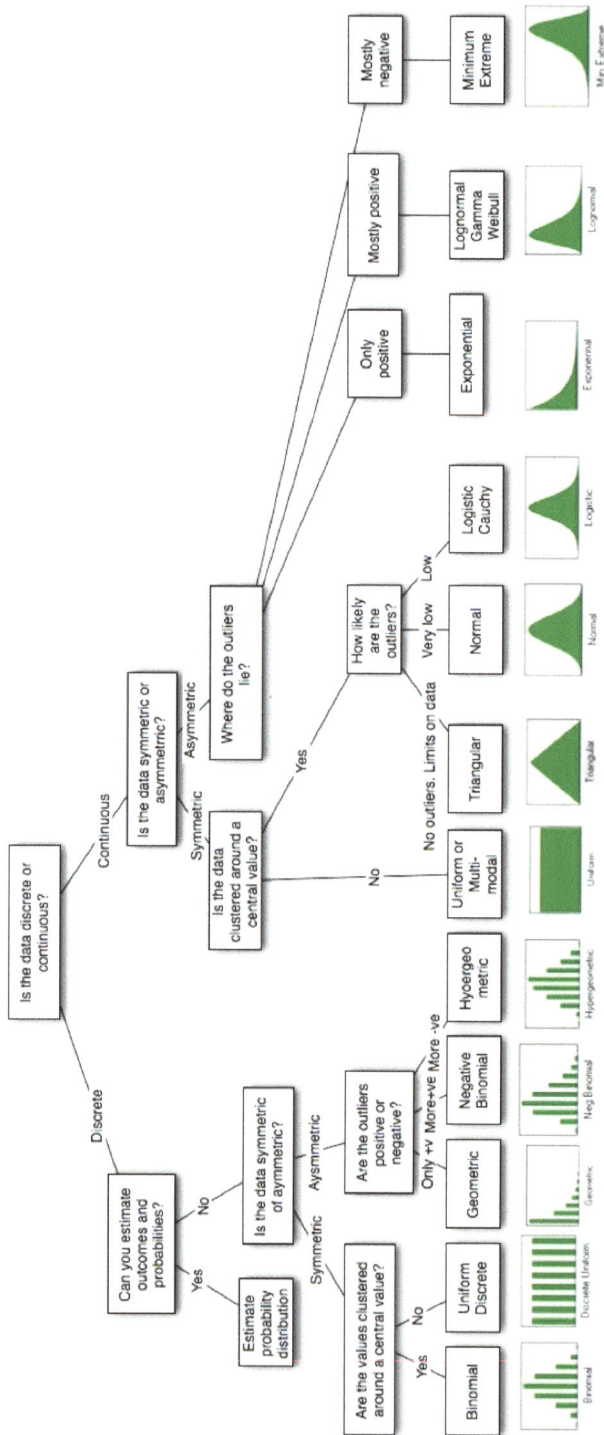

Figure II-8. Algorithm for defining the type of theoretical distribution depending on the problem being solved (A. Damodaran, 2011)

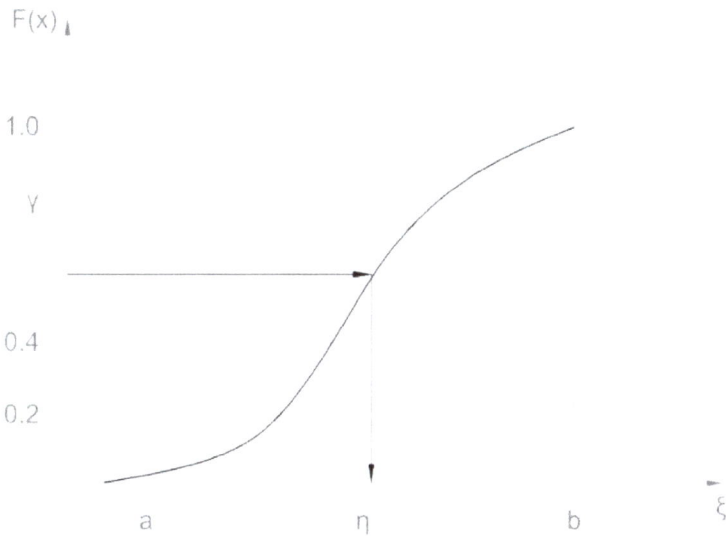

Figure II-9 Obtaining the values of a random quantity by means of cumulative distribution

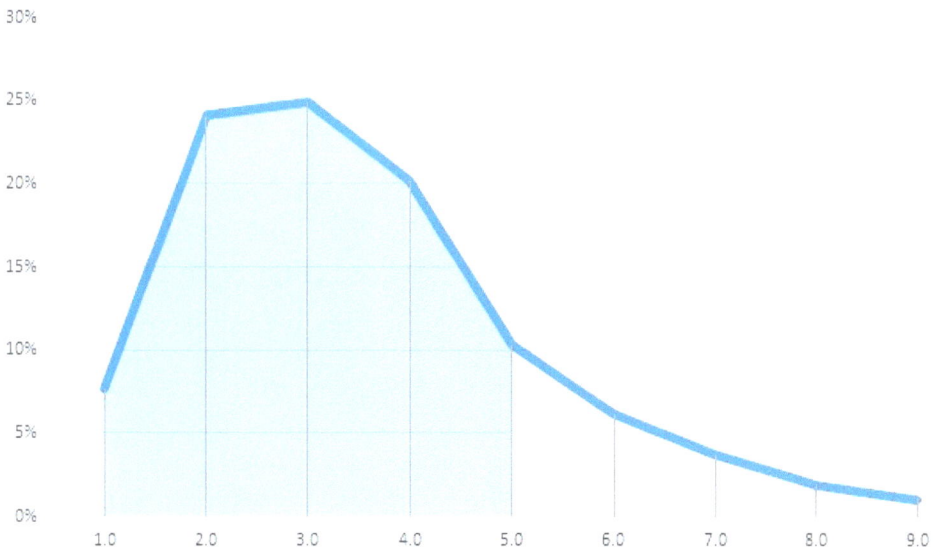

Figure II-10. Run along the entire range

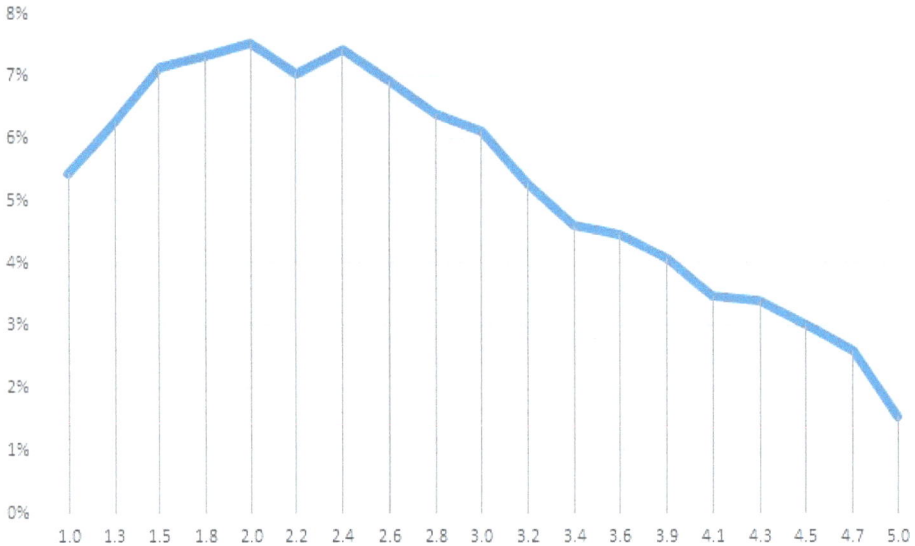

Figure II-11. Run in the range marked on Figure II-10

In some distributions, however, the grouping in equal intervals is not informative because considerable number of data falls in certain groups and there is a possibility of missing local extremums. In such case it is appropriate to use the famous approach of cluster analysis where the grouping is made depending on the results.

The problem is solved iteratively. By the first step the frequency of results in the entire range $[X_{min}; X_{max}]$ is determined (Figure II-10). In the second step, only the results in pre-determined one or more ranges are deduced (Figure II-11).

The proposed solution is not quite elegant but in our opinion it is sufficiently effective.

II.5. RISK

In millennial human history the genetically embedded reaction to unforeseen, life- or living condition-threatening situations has improved alongside the development of man's rational action. In the long process of the struggle for survival and for future successful development the change from intuitive perception of the existence of risk towards taking action to oppose the accompanying threats, has gradually taken clearer outlines as a characteristic of the possible manifestations and reactions in different fields of reality. The achievements of technologies in the new, technogenic and information society reveal greater opportunities for assessment and forecasting of risk situations until we get to the contemporary notions of identification and management of the impacts of risk. This has necessitated the creation and permanent improvement of the methodological and methodical tools for model binding of risk manifestations in different areas of their occurrence.

By its semantic meaning the concept of "risk" means danger, uncertainty and unknown outcome of an action taken. The most commonly used definition of the economic nature of risk it is examined as a possibility of negative deviation from the expected and actual results.

Risk identification, limitation and counteraction is an integral part of the process of decision making under the conditions of definiteness, indeterminateness, risk and counteraction.

The sources of risk are caused by external and internal causes. External causes are the result of changes to the macro-environment of organization's functioning and their manifestation is economic, strategic, financial, political and social in nature. Internal causes for the emergence of risk may be due to changes to the state of the enterprise such as technological and resource security.

The risk is manifested upon the concurrence of provoking events and leads to undesired consequences: technological, environmental, economic, to social tensions, loss in business, etc. Moreover, a definite risk "triggers" the emergence of other risks which can be interrelated with a multicolinear interaction or independent whereupon one or more danger flows of unforeseeable effect are formed. Their interrelated reflection as a whole causes the emergence and the manifestation of the integrated risk, i.e. the integrated risk in its nature is a unidirectional summary (overlayered) manifestation of destructive flows of danger seen as a generalized and undesired event.

Depending on the field of functioning of the managed system different approaches, methods, and means of decision-making in risk situations apply. This is related to the degree of indeterminateness and of a real, visible environment of factors and conditions of determinateness as possibilities of manifestation.

The size of the risk can be expressed by the mathematical expectation of its manifestation as average values of the impact of the factors on risk indicators, or by the mathematical probability of loss (harm) or damages it may cause. Logical and statistical methods of analysis and assessment are used in the processing of such information, mostly by means of mathematical expectation of the average value with the probability of its manifestation and of the statistical characteristics of variations or changes to the expected outcome.

In risk theory there are many definitions of its essence depending on the direction of scientific research and their applicability in practice (M. Dimov, 2015). The main thing that brings them together is that as a phenomenon risk is a non-coincidence of expectations and reality. From this point of view the most general idea about the quantity of the risk is the size or the value of this non-coincidence in a quantitative (as far as possible) or qualitative (as far as indicative) measure. Theoretical research and their actual corroborations in reality enable another peculiarity of risk manifestations to be revealed: they are most often causally linked, even where they are primarily registered. This explains the complexity upon establishing, analyzing the emergence and the quest for solutions for the essence of risk phenomena and processes and their overcoming.

By its semantic meaning the concept of "risk" means danger, uncertainty and unknown outcome of an action taken. In the science of economics (M. Dimov, 2015) there have been distinguished two theories of the essence of risk: classical and neoclassical. According to the classical theory the risk is measured by the mathematical expectation of the loss or harm that may be sustained by the occurrence of an unexpected event or in choosing a strategy for development of an activity (John Stuart Mill). The neoclassical theory interprets the concept of "economic risk", the solutions for which are subordinated to the concept of marginal utility, i.e. the guaranteed profit or the higher-than-expected utility (A. Marshall, J. M. Keynes). Keynes has an original definition of risk, which is presented in general by the analytical expression for assessment of the quantity of the risk K, as a non-coincidence of an expected utility A with permissible expenses E for the attainment thereof If p is the probability of a positive result, q is the probability that it cannot be attained $(p + q = 1)$, and $E = pA$ is the mathematical expectation that it can be attained, the risk K will be

$$K = p(A - E) = p(1 - p)A = pqA = qE. \qquad (II.11)$$

In this sense is the most commonly used definition of the economic essence of risk as a possibility of a negative deviation between the expected and actual results (M. Dimov, 2015).

II.6. NUMBER OF EXPERIMENTS

The problem with computational resource accompanies and, for some time, also limits the application of the Monte Carlo method. This problem is partially solved by the increase in contemporary computers' computation capacity. However, it should not be ignored, especially in solving problems of high degree of indeterminateness where the researcher has only an approximate idea of the nature of results obtained.

In addition to computational aspect the programmed number of runs (simulations) also has certain influence on the representativeness and reliability of results obtained.

Two types of deviations are possible: the number of runs is not sufficient, or such number is considerably greater than necessary. The negative consequences in both cases are different.

In case of insufficient number of runs there is a threat of not covering some of the so-called "low probability" events. As a rule this class of events is not treated by the classical mathematical statistics and thus the fact that they are covered by the Monte Carlo method is pointed out as its advantage that should not be ignored.

In case of greater number of runs there is a probability of the occurrence of the period effect in the sequence of quasirandom numbers generated by computer programs. Then, the probability of some events will be overestimated due to the periodicity of occurrence of quasirandom numbers. Thus, the number of runs must be set in conformity with the results of the test of the recurrence of the respective program generator.

E. S. Wentzel (1982) uses the apparatus of the classical mathematical statistics by linking the number of necessary data with the accuracy of application of the main statistical parameters.

1. If as a result of N realizations the frequency of occurrence of the event A is p^*, then the probability of occurrence of the event A (with a negligible deviation!) is within the range:

$$p^* \pm 2\sqrt{\frac{p^*(1-p^*)}{N}}. \qquad (II.12)$$

If we assume that the error in determination of the probability of the event A is not greater than Δ, it is necessary

$$N \geq \frac{4p^*(1-p^*)}{\Delta^2}. \qquad (II.13)$$

If $p^* = 0.30$ and $\Delta_p = 0.01$, then $N \approx 8\,400$.

2. If as a result of N realizations the value $\overline{x^*}$ of the variable quantity x is determined, the average value will be within the range

$$\overline{x^*} \pm \frac{2}{\sqrt{N}}\sqrt{\frac{1}{N}\sum_{i=1}^{N} x_i^2 - (\overline{x^*})^2}, \qquad (II.14)$$

And hence it follows

$$N \geq \frac{4\sigma_x^2}{\Delta^2}.$$

If $\Delta = 0.01, \sigma_x^2 = 0.1$, then $N \geq 400$.

On the basis of the main idea behind the Monte Carlo method, G. Trapov (2012) developed a methodology for assessment of the error made. Most often, the approximate value $\hat{a}$ of the indication being studied is obtained as an arithmetic mean of the sample

$$\hat{a} = \frac{\xi_1 + \xi_2 + \xi_3 + \cdots + \xi_N}{N}, \qquad (\text{II.15})$$

from N independent observations of the variable quantity ξ, whose mathematical expectation is $E\xi = a$.

Here, the assumption is that the quantity ξ has ultimate dispersion, i.e.

$$\sigma^2 = D\xi = E(\xi - a)^2 < +\infty.$$

When calculating the reliability $a, (0 \leq a \leq 1)$ of the assessment $\hat{a}$ with an accuracy ε the following equation is used

$$P\{|\hat{a} - a| < \varepsilon\} = 2\Phi\left(\frac{\varepsilon\sqrt{N}}{\sigma}\right) \approx a, \qquad (\text{II.16})$$

where $\Phi(x)$ is the function of Laplace.

Therefore, the reliability of assessment with an accuracy ε is

$$a \approx 2\Phi\left(\frac{\varepsilon\sqrt{N}}{\sigma}\right), \qquad (\text{II.17})$$

whereupon the approximate value of assessment's reliability is calculated in a given number of tests N, desired accuracy ε and certain dispersion σ^2 of the random quantity realized in the experiment.

With the set dispersion σ^2, the approximate number of tests N to be performed in order to obtain an assessment with an accuracy ε and desired reliability a. is established by formula (II.17). To that end we solve

$$\Phi(t_a) = \frac{a}{2},$$

and then from the equation

$$\frac{\varepsilon\sqrt{N}}{\sigma} = t_a,$$

we establish

$$N = \frac{\sigma^2 t_a{}^2}{\varepsilon^2}. \tag{II.18}$$

From the formula (II.18) it follows that in case of smaller dispersion the number of tests necessary to attain the defined reliability is lower.

In the conclusion it is assumed that the dispersion σ^2 of the indicator being studied is known. Unfortunately, when solving most problems this condition is not fulfilled. Then N_1 tests are performed and reduced dispersion is used for assessment of dispersion

$$\hat{\sigma}^2 = \frac{1}{N_1 - 1}\sum_{i=1}^{N_1}(\xi_i - \hat{a})^2.$$

Under (II.18) the approximate value N^* is computed for N. Then if $N^* > N_1$, $N^* - N_1$ experiments more are made, a reassessment is made for N and so on. Thus, after a finite number of steps we find the desired number of tests.

II.7. QUASI-MONTE CARLO METHODS [15]

There are different variants of the Monte Carlo method, which aim to accelerate and/or optimize the calculations made by using the classical method. Most often, the replacement of pseudorandom numbers by chains of random sequences is underlying for these solutions.

---

15 For additional information as regards the quasi-Monte Carlo methods, we recomend [20] Lemieux, C. (2009). Monte Carlo and Quasi-Monte Carlo Sampling, Springer. and [17] Kroese, D., T. Taimre, Z. Botev (2011). Handbook of Monte Carlo Methods, Wiley.

The main problem of the Monte Carlo method is the slow convergence, which usually is proportionate to the square root of the time for computations.

One approach to improve the convergence of the Monte Carlo method is the use of sequences of numbers from a limited interval (quasirandom sequences), instead of the generated pseudorandom numbers. The methods based on quasirandom sequences are called quasi-Monte Carlo methods (H. Niederreiter, 1992).

We will illustrate the two solutions with an example.

A square with side 1 (Figure II-12) is given. The area of the ABC triangle with known coordinates of vertices should be found.

The algorithm is not a complex one. The square is filled with points with coordinates $0 \leq X \leq 1$ and $0 \leq Y \leq 1$ (M. Mazhdrakov, D. Benov, 2010). Part of these points fall within the ABC triangle. Then

$$P_\Delta = \frac{N_\Delta}{N} \times 1, \qquad (II.19)$$

where N is the number of all points;

N_Δ - is the number of points falling into the ABC triangle.

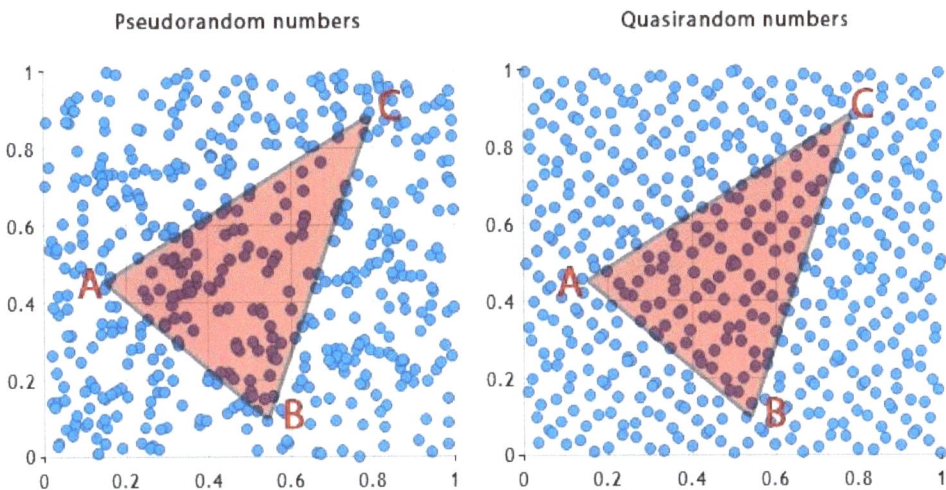

Figure II-12. Pseudorandom and quasirandom numbers

The generated pseudorandom numbers – the coordinates X and Y, are points of random distribution, i.e. with different distance from one another. Then, each point corresponds to different area and there might be "empty" zones. In order to reduce that effect but without any certainty that its influence will be eliminated the number of points must be increased, which however can result in recurrence of random numbers and hence an ambiguous determination of the area.

The generated quasirandom sequences are subject to certain conditions, for example: uniform distribution in a sufficiently small interval $d \ll 1$. Then each of the numbers falling within the triangle corresponds to an area d^2 and the area of ABC triangle will be

$$P_\Delta = N_\Delta d^2. \qquad (II.20)$$

The difference between the two approaches is essential when computing the volume of a body defined with a topographic surface (Figure II-13).

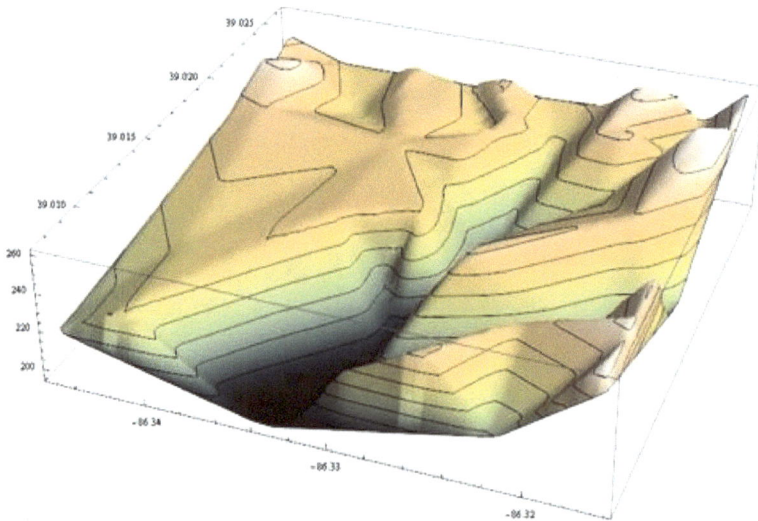

Figure II-13. Body limited by a topographic surface

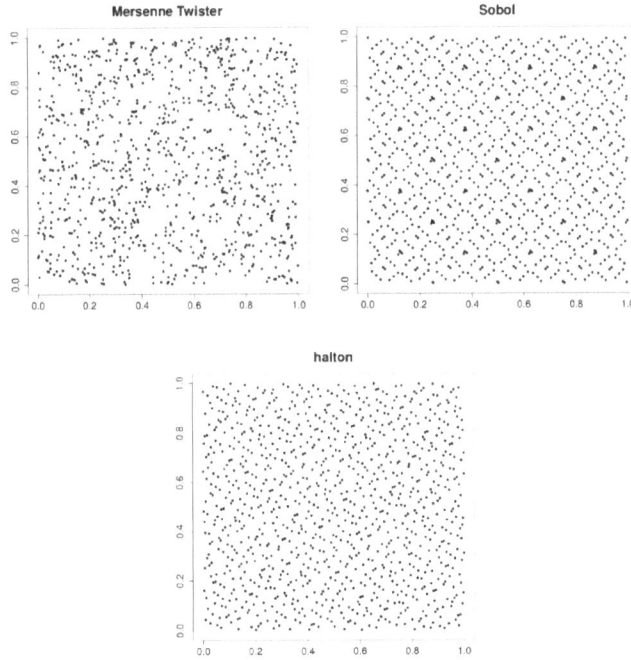

Figure II-14. A comparison of Mersenne Twister, Sobol and Halton

In both cases the replacement of the pseudorandom numbers with determined sequences of numbers leads to an acceleration of convergence and an increase of the accuracy of solution.

Of course, the quasirandom sequences used play the most important role in quasi-Monte Carlo methods. What matters are both the parameters of their distribution and the speed of generation.

Among the most popular quasirandom sequences are the ones of Sobol, J. Halton (1964) and Faure.

There are also other algorithms based on the Monte Carlo method (C. Lemieux, 2009; S. Brooks, A. Gelman, G. Jones et al., 2011).

II.8. SOFTWARE IMPLEMENTATION

Figure II-15 shows clearly enough (but not quite fully and accurately) a diagram of the practical implementation of the Monte Carlo method – generation of input data with which the mathematical model of the process (object) being studied is "enacted" and as a result certain distribution is obtained.

From a pragmatic point of view, the Monte Carlo method has two crucial advantages.

1. It is based on a simple algorithm that is relatively independent of the function being modeled.

2. It makes use of the strongest features of computers of last generations: their computing capabilities. Moreover, the need of considerable computational resources was pointed out as early as in the first publications about the Monte Carlo method.

Currently, there are many programs developed for the application of the Monte Carlo method in different fields of science and practice, which differ in terms of scope and opportunities.

Figure II-16 shows a generalized flowchart for realization of the Monte Carlo method. The main units are: 0 – organization of program's operation; 1 – random number generator; 2 – distribution of independent variables; 3 – generated massif of input data for a one-off single computation; 4 – module modeling the process being studied; 5 – organization of exit; 6 – analysis of results.

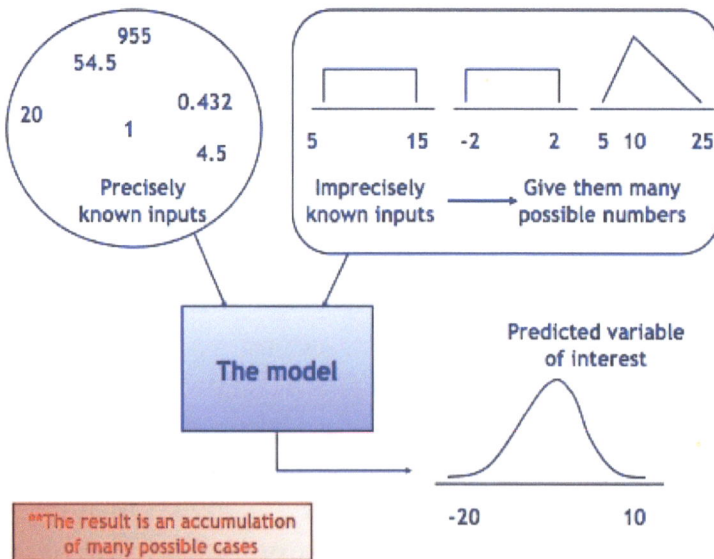

Figure II-15. Scheme for implementation of the Monte Carlo method

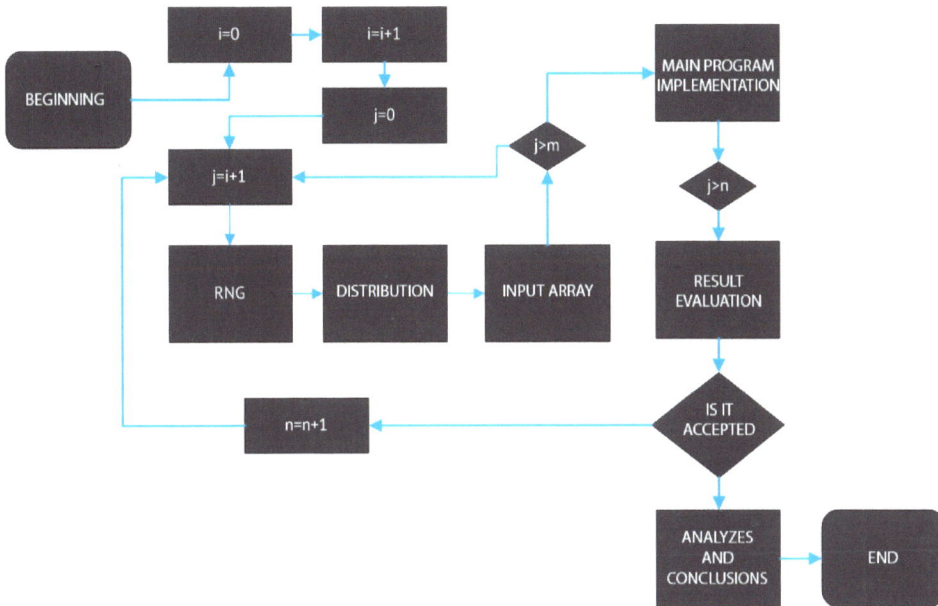

Figure II-16. Generalized flowchart for realization of the method

In program realization of the Monte Carlo method the problem is solved in considerable number of combinations of the values of input data. Therefore, there is certain probability of the occurrence of two events.

1. The computed results should not correspond to the physical essence of the process, and

2. Interruptions in the operation of computer program must be obtained.

The solving of the first problem is the task of the specialist user as it is directly linked to the input parameters entered by him.

The adverse events in the second problem must, however, be foreseen and accounted for in the software.

Most generally, "precautions" must be taken to avoid:

- inadmissible mathematical actions such as division by 0, even root of negative value, $\log N$, when $N \leq 0$, etc.;

- the computed value going beyond the permissible interval, e.g. a probability or relative frequency beyond $[0; 1]$;

- obtaining mathematically inadmissible values, e.g. $|\sin \alpha| > 1$;

- other errors typical for computer computations such as going beyond the announced dimensions of the massifs, obtaining indeterminate values of variables and others suchlike.

The programmer must also take into account the specific difficulties such as:

- forecasting of the duration of the run (see "*Number of experiments*");

- the assessment of the limits of solution sought.

We will try to present a Monte Carlo simulation by means of a pseudo-code.

```
//Setting values or computing parameters,
//which can be accurately determined
set A1=5.0;
set A2=Func(A1);

//Number of trials
set n=100000;

// Main simulation loop
for i=1,2,...,n
        //Setting values of parameters, which can not
        //be accurately determined

        //Generating a random number
        N=Random.Next();
        //Setting a value of the parameter on the basis of
        //a random number and distribution
        generate Z1=Distribution.Gauss(N);

        //A random number
        N=Random.Next();
        //Setting a value of the parameter on the basis of
        //a random number and distribution
        generate Z2=Distribution.Normal(N);

        //Computations with the participation of all parameters and
        // obtaining result R
```

```
      R[i]=Func(A1,A2,Z1,Z2);

//Determining the average value of the sought
//quantity and the standard deviation
...
//Determining the probable occurrence of
//a group of values
...
```

The average value of the sought quantity is

$$\hat{v} = \frac{1}{n}\left(\sum_{i=1}^{n} R_i\right),$$

and the standart deviation –

$$S.E. = \sqrt{\frac{1}{n-1}\left(\sum_{i=1}^{n} R_i^2 - n\hat{v}^2\right)}.$$

It is purposeful to group the results in sufficiently small intervals and to compute the probability of obtaining a value in the respective interval by the formula

$$\hat{v}_{x-y} = \frac{1}{n}\left(\sum_{i=1}^{n} f_{x-y}[R_i]\right) \times 100,\%,$$

where x and y are the limits of the interval.

Let us examine a more specific example, again in a pseudo-code.

We have to make a map of noise based on points of calculation situated on the terrain. We are interested in the most probable value of noise level in intervals of 3 dB(A).

Determining the noise level depends on the characteristic of sound source (motorcar traffic), the distance from the point of calculation and the relief, obstacles on the way of the sound wave, respectively.

The parameters that we could determine accurately are the number of traffic lanes and the distance between the source and the point of calculation.

The parameters to be modeled are traffic speed and intensity.

The procedure looks as follows.

```
//Setting values or computing parameters,
//which can be accurately determined
//Number of traffic lanes
set Ln=4;
//Existence of controlled intersections
set RI=false;
...

//Number of trials
set n=100000;

//Main simulation loop
for i=1,2,...,n
        //Setting values of parameters, which can not
        //be accurately determined

        //Generating a random number
        N=Random.Next();
        //Setting a value of the parameter on the basis of
        //a random number and distribution
        //Traffic speed
        generate V=Distribution.Normal(N);

        //Generating a random number
        N=Random.Next();
        //Setting a value of the parameter on the basis of
        //a random number and distribution
        //Number of motorcars in one lane
        generate AC=Distribution.Normal(N);
        ...

        //Computations with the participation of all parameters and
        //obtaining result R
        R[i]=Func(Ln,RI,...,V,AC,...);

//Determining the average value of the sought
//quantity and the standard deviation
```

```
...
//Determining the probable occurrence of
//a group of values
...
```

In order to determine the most probable level of noise in intervals of 3 dB(A), we must determine the limits of intervals. At a base noise level of 45 dB(A) and upper limit $\min(\max L_A) > 80$ dB(A), the intervals will be $< 45, [45; 48), [48; 51), [51; 54), \cdots, [78; 81), \geq 81$ dB(A).

In order to determine the probability of noise level in a specific point of calculation being in the interval $[69; 72)$ dB(A), we must determine the number of results falling within the set interval (n_{69-72}) and divide it by the total number of experiments (n)

$$L_{A,69-72} = \frac{1}{n}(n_{69-72}) \times 100, \%.$$

REFERENCES

[1] AP (2015). Former lottery official sentenced for trying to rig $14M win. CBS News. Des Moines, Iowa.

[2] Brooks, S., A. Gelman, G. Jones, et al. (2011). Handbook of Markov Chain Monte Carlo, Chapman & Hall.

[3] Cruz, J. (2006). "Implementing Five PRNG Algorithms in Cocoa." MACTECH 22(3).

[4] Damodaran, A. (2011). Probabilistic approaches to risk: 61.

[5] Dimov, L. (1979). Mine Surveying Handbook. Sofia, Tehnika [in Bulgarian].

[6] Dimov, M. (2015). Managerial aspects of integrated risk in the mining and quarrying industry. Underground Contruction. Sofia, MGU St. Ivan Rilski. Doctor: 55 [in Bulgarian].

[7] Eadie, W., D. Drijard, F. James, et al. (1971). Statistical Methods in Experimental Physics. Amsterdam.

[8] Entacher, K. (1997). A collection of selected pseudorandom number generators with linear structures, Austrian Science Foundation: 25.

[9] Forsythe, G. E., M. A. Malcolm, C. B. Moler (1977). Computer methods for mathematical computations, Prentice-Hall.

[10] Gentle, J. (2005). Random number generation and Monte Carlo methods, Springer.

[11] Halton, J. (1964). "Algorithm 247: Radical-inverse quasi-random point sequence." ACM: 701.

[12] Hassan, A. R., J. C. Jimenez (2014). Which team will win the 2014 FIFA World Cup?, Universidad EAFIT.

[13] Ivanova, V. (1984). Random Numbers and Applications. Moscow, Finance and Statistics [in Russian].

[14] Kendall, M., B. Babington Smith (1938). "Tracts for Computers." Royal Statistical Society.

[15] Kendall, M., B. Babington Smith (1939). Tables of random sampling numbers. London, Cambridge University Press.

[16] Knuth, D. (1968). The Art of Computer Programming. United States, Addison-Wesley.

[17] Kroese, D., T. Taimre, Z. Botev (2011). Handbook of Monte Carlo Methods, Wiley.

[18] Kumar, M. (2011). Quantum: Einstein, Bohr, and the Great Debate about the Nature of Reality, W. W. Norton & Company.

[19] Lahvicka, J. (2012) Using Monte Carlo simulation to calculate match importance: the case of English Premier League Munich Personal RePEc.

[20] Lemieux, C. (2009). Monte Carlo and Quasi-Monte Carlo Sampling, Springer.

[21] Malone, M. (2015). TIFU by using Math.random(). Medium. US, The Medium Corporation. 2016.

[22] Matsumoto, M., T. Nishimura (1998). "Mersenne twister - a 623-dimensionally equidistributed uniform pseudo-random number generator." ACM Transactions on Modeling and Computer Simulation 8(1): 3-30.

[23] Mazhdrakov, M., D. Benov (2010). Algorithm and software for area division. Geomedia. Sofia: 30-31 [in Bulgarian].

[24] Niederreiter, H. (1992). Random Number Generation and Quasi Monte Carlo Methods. Philadelphia, Pennsylvania, USA.

[25] Obretenov, A. (1978). Probability and statistical methods. Sofia, Nauka i Izkustvo [in Bulgarian].

[26] Regius, H. (1536). Utrisque Arithmetices Epitome.

[27] Saito, M., M. Matsumoto (2006). <u>A Uniform Real Random Number Generator Obeying the IEEE 754 Format Using an Affine Transition</u>. International Conference on Monte Carlo and Quasi-Monte Carlo Methods in Scientific Computing.

[28] Schutt, R., C. O'Neil (2014). <u>Doing Data Science</u>. Sebastopol, CA, O'Reilly.

[29] Shreider, Y. A. (2014). <u>The Monte Carlo Method: The Method of Statistical Trials</u>, Elsevier Science.

[30] Smirnov, N., I. Dunin-Barkovskii (1962). <u>Course of probability theory and mathematical statistics</u>. Moscow, Nauka [in Russian].

[31] Sobol, I. M. (1994). <u>A Primer for the Monte Carlo Method</u>, Taylor & Francis.

[32] The Guardian (2016). Sepp Blatter: European competition draw was rigged by cooling balls <u>The Guardian</u>. London, UK.

[33] The RAND Corporation (1955). <u>A Million Random Digits with 100,000 Normal Derivates</u>, The Free Press.

[34] Tippett, L. H. C. (1927). <u>Random Sampling Numbers</u>. London, CUP.

[35] Trapov, G. (2012). Application of statistical modeling for the solving of mineral extraction tasks. Sofia, MGU St. Ivan Rilski. Doctor [in Bulgarian].

[36] University of Notre Dame (2006). "Understanding Randomness in Biology." from www3.nd.edu.

[37] von Neumann, J. (1951). "Various Techniques Used in Connection With Random Numbers." <u>Monte Carlo Method, National Bureau of Standards Applied Mathematics Series</u> 12: 36-38.

[38] Wentzel, E. S. (1982). <u>Probability Theory (first Steps)</u>, Mir Publishers.

SECTION III.
MONTE CARLO AND...

M. Mazhdrakov, D. Benov, D. Benova

III.1. ... THE NUMBER π

Pi (π) is a mathematical constant, the ratio of a circle's circumference to its diameter and it has universal application. Being an irrational number π cannot be expressed as a fraction of two integers, such as 22/7. This was proven by Johann Heinrich Lambert in 1761. Also, π is a transcendental number (proven by Ferdinand von Lindemann in 1882), that is, it is not the root of any polynomial with rational coefficients. Due to the transcendence of π it is a non-constructible number. From the requirement that coordinates of all points that can be constructed by using a compass and straightedge must be constructible numbers it follows that it is impossible to square the circle (to construct a square with the same area as a given circle by using only compass and straightedge) (J. L. Berggren, J. M. Borwein, P. Borwein, 2004).

Fractional numbers that have been used to determine π throughout the years are $\frac{22}{7}, \frac{333}{106}, \frac{355}{113}, \frac{52163}{16604}, \frac{103993}{33102}, \frac{245850922}{78256779}$ (P. Eymard, J. P. Lafon, 1999).

A detailed bibliography of publications related to the approximation of π was published by N. Beebe (2014).

The Pi Day was proposed in 1988 by Larry Shaw, a physicist at Exploratorium Museum in San Francisco. He used the US month/date format: 3/14 at 1:59 o'clock, which coincides with the first digits of the number π = 3.14159... The US House of Representatives supported the celebration of the holiday.

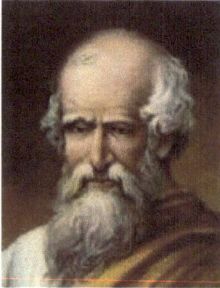

Archimedes (287-212 BC) developed the method of polygon to determine π

Isaac Newton (1643-1727) used infinite series to compute a 15 digit **approximation of π**

Leonhard Euler (1707-1783) popularized the use of the Greek letter π to denote the ratio of a circle's circumference to its diameter

John von Neumann (1903-1957) is part of the team that first used electronic computer to approximate π

Since the fraction 22/7 is a common approximation of π, Pi Approximation Day will be July 22 (22/7 in the day/month date format).

In this chapter we will examine some algorithms for approximation of π by using the Monte Carlo method.

Buffon's Needle. It is accepted that the first version of the "Monte Carlo" idea is the famous experiment "Buffon's Needle" of 1777. (G.-L. Buffon, 1777; E. Siniksaran, 2008).

The classic formulation of the problem is to drop a needle 1 inch (2.54 cm) ling on a surface made of parallel strips at a distance of 2 inches (5.08 cm) from one another (Figure III-1). If the needle lies across a line it scores a hit (needle a, Figure III-1). Buffon's hypothesis is that the ratio of all attempts to the number of falls tends to the number π

$$\pi \approx \frac{2Ln}{dh},\tag{III.1}$$

where L is needle's length (Figure III-1);

d – the distance between the lines (Figure III-1);

n – the number of all attempts;

h - the number of hits.

Using the Monte Carlo method to simulate "the fall" of the needle we made an experiment by "throwing" the needle different number of times.

The throw of the needle on the parallel lines is modeled by random values of the coordinates of one end of the needle − X_A, Y_A, and the needle's pointing angle (Figure III-2). The coordinates of the other end of the needle − X_B, Y_B, are computed and then it is checked if it crosses any of the parallel lines and the crossing is counted as a hit.

The results obtained are shown in Figure III-3.

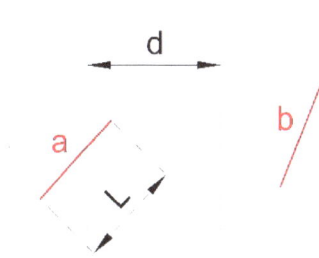

Figure III-1. Formulation of the Buffon's Needle problem

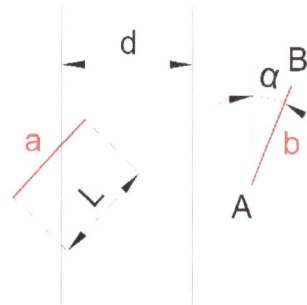

Figure III-2. Modelling of the experiment

Von Neumann's Method. The approximation of π by von Neumann method is made as per the following scheme. We have a square and a circle with side and radius of 1 (Figure III-4). We compute the coordinates of an arbitrary point in the square by using random numbers and we check if this point falls within the circle or lies outside

$$P(x^2 + y^2 < 1) = \frac{A_{cir}}{A_{sq}} = \frac{\pi}{4}. \qquad (III.2)$$

A program realization of C# is given on Listing III-1.

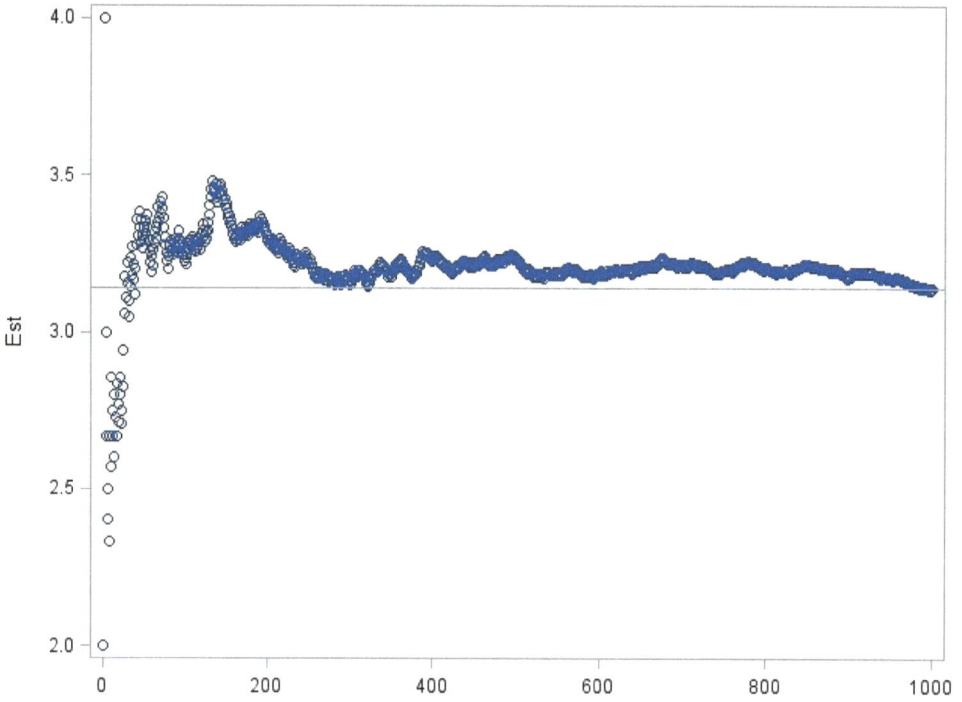

Figure III-3. Buffon's Needle: computed values of π

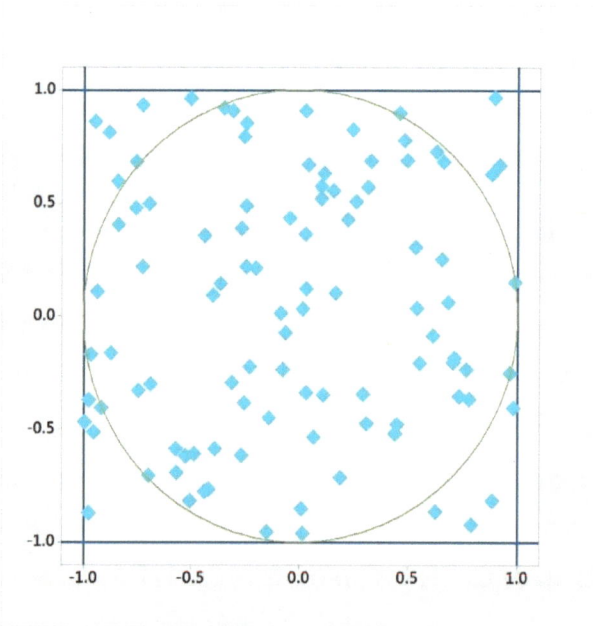

Figure III-4. Formulation of the problem of approximation of π by von Neumann's method (M. Paret, 2015)

Listing III-1

```
1.    int Simulations = 1000;
2.    int InTheCircle = 0;
3.    for (int i = 1; i <= Simulations; i++)
4.    {
5.        double X = Rand(1);
6.        double Y = Rand(1);
7.        if (Dist(0, 0, X, Y) <= 1)
8.            InTheCircle++;
9.    }
10.
11.   double pi = 4 * InTheCircle / Simulations;
```

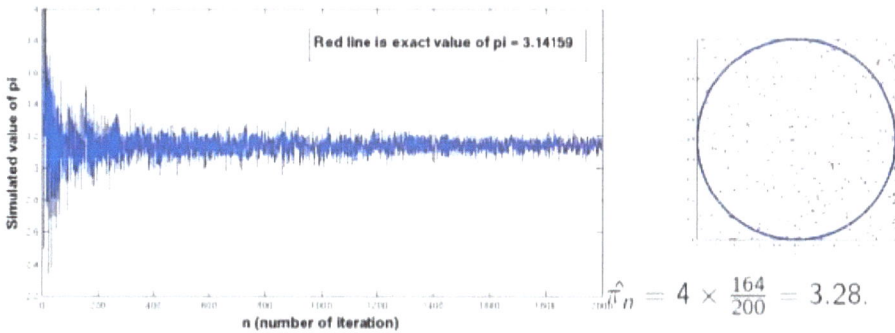

$$\hat{\pi}_n = 4 \times \frac{164}{200} = 3.28.$$

Figure III-5. Values of π calculated by von Neumann's method

$\pi =$ 3.14159	26535	89793	23846	26433	83279	50288	41971	69399	37510
58209	74944	59230	78164	06286	20899	86280	34825	34211	70679
82148	08651	32823	06647	09384	46095	50582	23172	53594	08128
48111	74502	84102	70193	85211	05559	64462	29489	54930	38196
44288	10975	66593	34461	28475	64823	37867	83165	27120	19091
45648	56692	34603	48610	45432	66482	13393	60726	02491	41273
72458	70066	06315	58817	48815	20920	96282	92540	91715	36436
78925	90360	01133	05305	48820	46652	13841	46951	94151	16094

Figure III-6. Excepts from the table with the result obtained for π (N. Metropolis, G. Reitwiesner, J. von Neumann, 1950)

The results of computations depending on the number of trials are given in Figure III-5. When increasing the number of iterations, the computed value increasingly approximates the real one.

By the described method N. Metropolis, G. Reitwiesner, J. von Neumann (1950) got π with 2,000 digits after the decimal mark (Figure III-6).

III.2. ... BLACK SWANS

When the crisis is over the most common question is "Why?" Why did XXXL Bank close at 16:59:59 o'clock on Tuesday as a symbol of stability and at 08:00:01 o'clock on Wednesday it announced that it stops payments? What factor can put a reputable financial institution operating for 150 years out of business?

The usual suspect is management, and, in particular: mistakes in Forecasting and Planning stages. Some events that either were not foreseen or had resulted in unforeseeable consequences have been realized.

In practice, the application of scientific methods in management started in the end of XIX-th century by the statistical models (let us recall why William S. Gosset published under the pen name Student). Developments in theoretical research and, in particular, the computing machinery have enriched the panoply of mathematical methods but the unwritten rule that they govern the model and not the actual object has remained.

In statistical processing of real data, results with values that considerably differ from the others are observed. There are no unequivocal and generally accepted criteria to establish the permissible differences in the values of separate results. In each field of human knowledge the approaches are specific, with considerable amount of subjectivism. As a rule, however, such phenomena are declared events of negligibly low probability of occurrence and are eliminated or "ex officio" get some value as it is assumed that the exclusion of considerably differing values does not have an impact on final outcome of processing.

Nassim Nicholas Taleb (2006) (N. N. Taleb, 2010) rejects this formulation and accepts that:

(1) events that are beyond the scope of customary expectations are realized quite often,

(2) catastrophic phenomena in the history of mankind are due to such events, and

(3) these events are explained postfactum.

N. N. Taleb (2006) calls the "low-probability" event "black swan" by analogy to the XVII-th century: deterministic saying "All swans are white". However, it turned out that in Australia there are black swans as established in 1864. In Bulgaria, there is a similar saying: the *Shop* in front of giraffe's cage in the zoo: "There is no such animal" but as it is known the giraffes live in Africa and black swans live in Australia.

According to N. N. Taleb the majority of scientific discoveries and historical events that appear unexpectedly and/or that are not linked to historical context are black swans. Examples for such black swans are the internet, personal computer, the First World War, and September 11. by analogy, in the present (June 2016) the decision of the referendum on United Kingdom's leaving the European Union that became known as 'Brexit' was declared to be a black swan.

N. N. Taleb opposes the sudden abrupt and destructive occurrence of Black Swans to the popular normal distribution, whose graph is often referred to as a "bell curve" and which implies a smooth limited and balanced change to the indicators.

In substance, N. N. Taleb (2006) points out the limited prognostic capabilities of probabilistic models, especially the ones based on normal distribution. By using statistical processing the majority of results obtained is presented one or two, or more, nut always a finite number, of synthesized parameters. For the normal distribution these are the average value of the quantity $\bar{x}$, and its average squared deviation σ_x. These two quantities produce the distribution of the probability of realization of one value of the indicator or another.

Of course, the normal distribution is not universal. It is deduced for processing of random errors in geodesic measurements.[16] (C. Gauss, 1809; P. Laplace, 1812), which is also visible on the banknote of 10 DM (Figure III-7), (C. Gauss, 1809; P. Laplace, 1812), which can also be seen on the 10 DM banknote (Figure III-7) depicting the triangulation of the Kingdom of Hanover and a sextant.[17]. We owe the term "normal" distribution to Karl Pearson. Inn early 20th century the normal distribution was set as an assessment of accuracy in industrial standards and hence was defined as a "standard" one or simply as "standard".

The application of the three sigma rule is justified in the processing of realized observations. If conditions are changed it is quite possible some results we have defined as being of "low probability: to become "predominant".

According to this rule we can expect that 1 in 370 cases will fall outside this interval. In cosmic studies the permissible deviation is 1 in 1000 cases, i.e. the rule is $\pm 3.3\sigma$.

Prognostic possibilities of normal distribution are limited for two reasons.

1. The values of $\bar{x}$ and σ_x are obtained under certain conditions and can be extrapolated only if the same conditions are preserved.

2. As real data are being processed it is accepted that the result of measurements must be limited within a predefined limits, most often in the interval $[\bar{x} - 3\sigma_x; \bar{x} + 3\sigma_x]$, also popular as the three sigma rule.

As aforesaid, the statistical distribution is a mathematical model of the frequency of occurrence of certain values of the indicator being studied as a function of a limited number of parameters whose values have been empirically established.

---

16 N. N. Taleb points out, not quite accurately, that this distribution refers to errors from computations; the errors from rounding in the case of computing with fixed decimal mark are distributed uniformly and are equally probable.

17 A device that measures the angle between the Sun or a star and a point on the horizon or that measures a horizontal angle between two objects (most often navigation reference points). The name comes from the Latin word *sextāns* meaning "one-sixth".

Figure III-7. 10 DM (deutsche mark) (c) Deutsche Bundesbank

When applying these models the following is implied:

— homogeneity of indicators;

— an interval of studying limited by objectively existing minimum and maximum values;

— elimination of the impact of systemic factors.

Where these conditions are not fulfilled some unexpected results may be obtained.

N. N. Taleb cites a curios example. He examines the distribution of the indicator of average height in the species of

Homo Sapiens − $\bar{x} = 167$ cm with dispersion $\sigma_x = 16.7$ cm and gets to some sensational (in his opinion) results. By changing the interval above $\pm 3\sigma_x$ N. N.

Taleb gets a difference of:

± 10 cm, i.e. height from 1.57 to 1.77 m − 1:6;

± 20 cm, i.e. height from 1.47 to 1.87 m − 1:44;

± 30 cm, i.e. height from 1.37 to 1.97 m − 1:740;

...

± 60 cm, i.e. height from 1.07 to 2.27 m − $1:10^9$.

These results are justified if we take into account that mankind is about 6.5×10^9 and currently the NBA players are included in this category. The next step ± 70 cm − is one of probability considerably lower than the population on Earth and, of course, it cannot be realized with present-day population.

Let us consider the case in more detail. First, we will note that the determination of the average height of the population is not an "academic" problem. The values obtained matter much more for garment producers, for manufacturers of aircraft, wagons and other suchlike, for determining the dimensions of operator's consoles, etc. Thus, anthropometric research is regularly conducted in all developed countries (K. Norton, T. Olds, A. S. Commission, 1996). Moreover, the following must be taken into account.

1. Human height depends on sex, age, race, lifestyle (eating habits, physical load, etc.).

2. Average height must be determined in the interval between the observed minimum and maximum values.

3. The research is meaningful only if it concerns individuals in whom the average height is stabilized (in Bulgaria: between 30 and 40 years of age).

4. Studies must refer individually to men and women; in Bulgaria the average height in men is 172 cm, and in women it is 158 cm.

If the above conditions are fulfilled "more real" results are obtained. If empirical data correspond to the normal distribution with a dispersion of ± 3 cm, then we should expect:

– for men $\bar{x} = 172$ cm, of which 68% fall within the interval 172±3 %, 95% fall within the interval 172±6, 99.7% fall within the interval 172±9 and 99.97% fall within the interval 172±12;

– for women $\bar{x} = 158$ cm, of which 68% fall within the interval 158±3, 95% fall within the interval 158±6, 99.7% fall within the interval 158±9, and 99.97% fall within the interval 158±12.

In our opinion the Monte Carlo method reacts adequately enough to the Black swan phenomenon whatever it means. In it the extreme values are not eliminated as all results obtained are made use of, especially where empirical distributions are used. Another advantage is that no possible results are eliminated; on the contrary, according to the law of large numbers the so-called "low probability" events are also included.

Compared to classical mathematical statistics the Monte Carlo method has greater possibilities for studying low probability events. In our opinion this is due to three reasons.

1. There is no limitation on the entrance. Input quantities can be set by empirical or theoretical distributions (the normal distribution is nt mandatory).

2. The number of runs is great enough and it includes combinations of input data that would be excluded in the classical processing.

3. The rich output information is the basis for a thorough analysis and forecasting of adverse phenomena.

III.3. … FUNCTIONALLY LITERATE STUDENTS

The Monte Carlo method is not included in secondary school curriculum. The experts from Brussels, however, suggest that the students must e familiar with this approach because one of the functional literacy tests (PISA[21]) (M. Lockheed, T. Prokic-Bruer, A. Shadrova, 2015) for 15-year olds includes at least one problem from the field of stochastic modeling.

Please pay attention to the image! There are two landscapes on the screen, both on terrains of markedly opposite slope. The first landscape is cheerful, with green grass and trees of well-manifested crowns. In the other one the grass has turned yellow and there is almost no high vegetation. A table with information about three factors which are different for each landscape is given. The values and the condition of the problem have been changed by the authors.

Table III-1. Values of factors for the two landscapes

Factor	Landscape A	Landscape B
1	240±30	170±35
2	1100±120	1250±250
3	63±5	98±9

The average value $\bar{x}$ and the average squared deviation $\pm\sigma_x$ are shown for each factor.

The problem is worded thus: What is the difference in the two landscapes due to?

The functionally literate person, after a brief reasoning, guesses that the problem is in the field of stochastic modeling and finds out that the Monte Carlo method should be employed[22].

Before proceeding with resolving the problem he assumes that:

(1) the factors are independent, and

(2) the values of indicators are normally distributed.

The differences between the said factors for the two landscapes are modeled and one sees which ones are essential and which are not.

---

[21] Programme for International Student Assessment.

[22] The problem can also be solved by dispersion analysis.

65
22 to 110

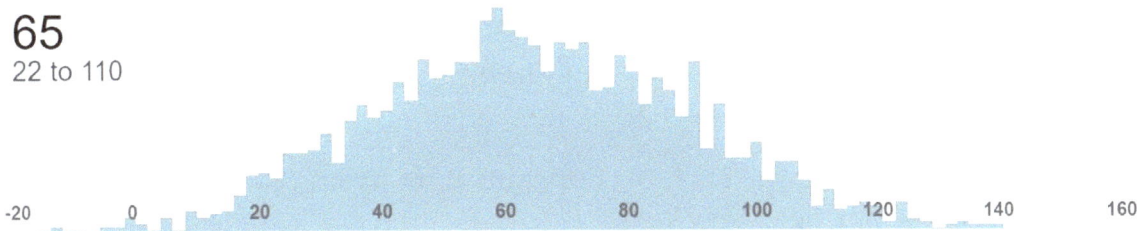

Figure III-8. Factor 1 – positive effect for landscape A

150
-110 to 420

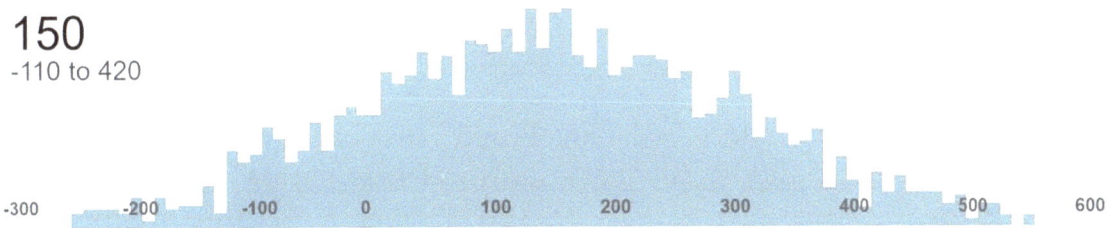

Figure III-9. Factor 2 – neutral

35
25 to 45

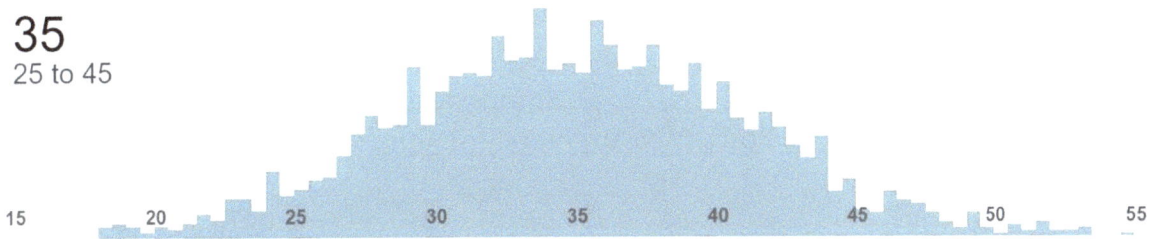

Figure III-10. Factor 3 – positive effect for landscape B

Under the said conditions most probably the effect of factor 1 is positive for landscape A, factor 3 is negative for landscape B and the factor 2 is neutral (Figure III-8, Figure III-9, Figure III-10).

REFERENCES

[1] Beebe, N. (2014). A Bibliography of Publications on the Numerical Calculation of π.

[2] Berggren, J. L., J. M. Borwein, P. Borwein (2004). Pi: A Source Book, Springer New York.

[3] Buffon, G.-L. (1777). "Histoire naturelle générale et particulière." 615.

[4] Eymard, P., J. P. Lafon (1999). The Number Pi, American Mathematical Society.

[5] Lockheed, M., T. Prokic-Bruer, A. Shadrova (2015). The Experience of Middle-Income Countries Participating in PISA 2000-2015.

[6] Metropolis, N., G. Reitwiesner, J. von Neumann (1950). "Statistical treatment of values of first 2,000 decimal digits of e and of π calculated on the ENIAC." Mathematical Tables and Other Aids to Computation 4(30): 109-111.

[7] Norton, K., T. Olds, A. S. Commission (1996). Anthropometrica: A Textbook of Body Measurement for Sports and Health Courses, UNSW Press.

[8] Paret, M. (2015). A Simple Guide to Using Monte Carlo Simulation to Estimate Pi. The Minitab Blog. 2016.

[9] Siniksaran, E. (2008). "Throwing Buffon's Needle with Mathematica." The Mathematica Journal 11(1): 71-90.

[10] Taleb, N. N. (2010). The Black Swan: Second Edition: The Impact of the Highly Improbable Fragility", Random House Publishing Group.

6

PART 2

ENGINEERING
applications

Actual and Predicted Graph

SECTION IV.
QUALITY MANAGEMENT IN EXTRACTION OF ORES AND MINERALS

M. Mazhdrakov, K. Boev, N. Valkanov, St. Topalov, D. Benov

The Monte Carlo method is very suitable for finding the optimum solution in processes characterized by complex relations between indicators. An example for that is quality management in extraction of ores and minerals (M. Mazhdrakov, D. Benov, 2006).

IV.1. EFFICIENCY

Quality indicators are among the most important criteria for management of extraction works. In particular, in extraction of non-ferrous metal ores the quality of ore extracted determines the technological extraction of the useful components in dressing. All researchers agree that stabilization of quality indicators has decisive importance for the efficiency of dressing process, which is expressed by the coefficient of technological extraction

$$\varepsilon = \frac{c_o - c_w}{c_c - c_w} \frac{c_c}{c_o},$$

where c_o is the useful component content in the ore, %;

c_c is the useful component content in the concentrate, %;

c_w is the useful component content in the waste, %.

Based on numerous studies sufficiently reliable correlations have been established between technological extraction and quality characteristics of ore flow.

K. Boev, I. Vaklev (1978) found a significant non-linear correlation between the changes to technological extraction of the useful component $\Delta\varepsilon$, and the deviation of its content from the scheduled one Δc_o.

$$\Delta\varepsilon = 8.88|\Delta c_o|^2 + 21.32|\Delta c_o| + 1.84 \qquad (IV.1)$$

It can be seen that the deviation in ore quality by 3 conditional units causes a difference in extraction by about 1% (Figure IV-1).

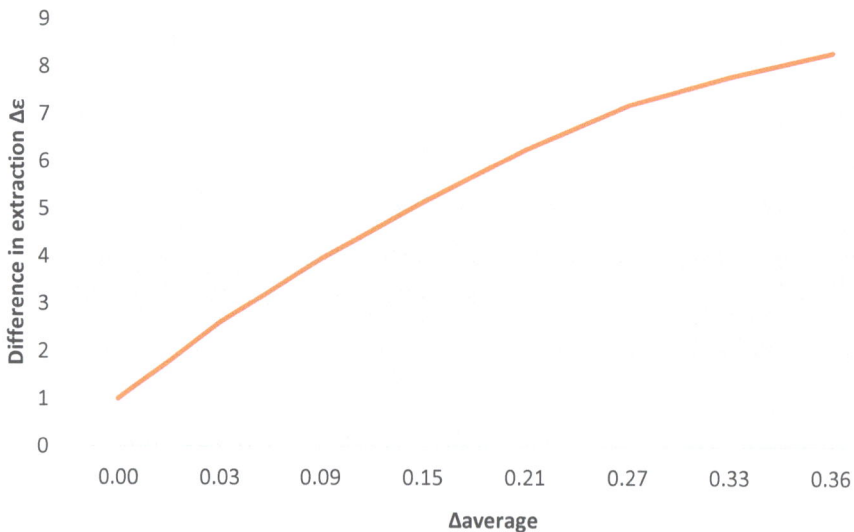

Figure IV-1 Metal content and extraction at the time when the studies were conducted were classified information and the authors have used conditional values of indicators.

Based on an analysis of literature and his own research I. Andonov (2003) reached the conclusion that the values of average content and standard deviation in the mineral raw material fed for processing have the main impact on the level of extraction of useful component in the concentrate, i.e.

$$\varepsilon = f(c_o, \sigma_o). \qquad (IV.2)$$

In order to determine the relation (IV.2) data about copper content in the raw material fed for processing, in the concentrate and in waste product were collected at each 2 hours for a period of 12 months. Figure IV-2 shows graphs of the function (IV.2) with different values of average metal content. It can be seen that when changing the standard deviation by 0.1, extraction drops by 5% on the average.

It is important that the increase in extraction of the useful component is achieved without special additional expenses, i.e. the additional concentrate obtained is "pure profit" to the amount of

$$\Delta P = \frac{Q_o \bar{C}_o}{\bar{C}_c} \Delta \varepsilon P_c, \text{USD,} \qquad (\text{IV.3})$$

where Q_o is the weight of ore processed, t;

$\bar{C}_o$ is the average useful component content in the ore, %;

$\bar{C}_c$ is the average useful component content in the concentrate, %;

$\Delta \varepsilon$ is the technological extraction in dressing;

P_c is the price per 1 t of concentrate in USD.

In case that 5 million t of ore are extracted with an average copper content of 0.4% and is processed into a concentrate with 21% content, if extraction is increased by 0.01, about 950 t of concentrate are produced. Then, if the price for 1 t of concentrate on the London Metal Exchange is $700, the positive effect is $665 thousand.

Extraction planning in case of target function of maintaining certain metal content in the raw material is a complex enough non-linear optimization problem, i.e. it is quite appropriate for the application of the Monte Carlo method.

K. Boev, I. Vaklev (1978) published "Method for planning of quality indicators upon ore extraction in underground mines" based on the Monte Carlo method. For monthly planning within one year the extraction blocks $i = 1,2,3,...,n$ are modeled by the period's productivity $Q_i \pm \Delta Q_i$ and metal content $c_i \pm \Delta c_i$. The values of these quantities are determined by statistical analysis of the results of mine's operations. In the modeling of the process the stated quantities change by using random numbers.

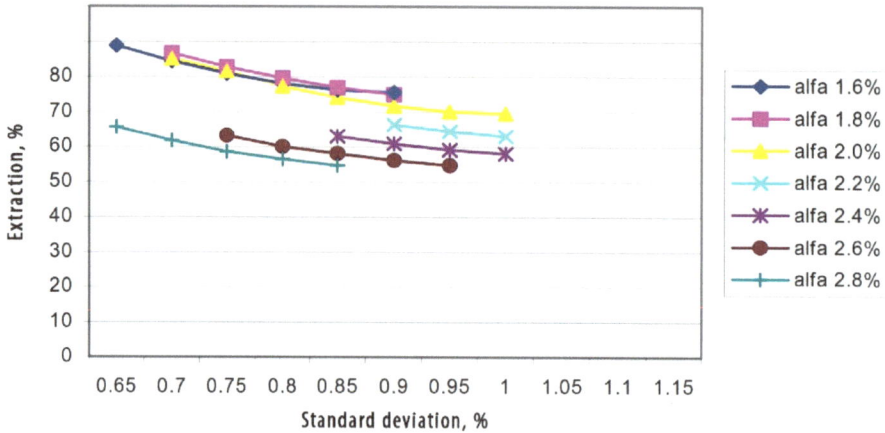

Figure IV-2. Graphs of the function $\varepsilon = f(\sigma_p)$, obtained by averaging the values of extraction (ε) and the standard deviation (σ_p), by using different methods to smooth the data from the "Second" aggregate

The level of wide-spread computing machines in Bulgaria in that period is not quite high. Thus, the authors used a table for the values of random numbers (N. Smirnov, I. Dunin-Barkovskii, 1962) and limited the number of simulations.

IV.2. MONTHLY PLANNING

Monthly planning implements the link between the annual plan for development of mining works and the current planning for a period of 1 to 10 shifts (M. Mazhdrakov, D. Benov, G. Trapov, 2013). It ensures the availability of industrial reserves in sufficient volumes and quality ready to be stripped thus enabling the trouble-free performance of mining works. Taking into account that the formation of quality indicators in ore extraction is a complex process it is appropriate to apply a probabilistic approach by using the Monte Carlo method. We will apply the method for the conditions of the mine stated in the publication under consideration.

The mathematical model for development of mining works in the month is reduced to the following. Extraction is ensured by n

workplaces (blocks) as for each of them the following has been established by statistical means:

- q_i is the average monthly productivity, t/month;

- dq_i is the average deviation of productivity, t/month;

- $\bar{c}_i$ is the average metal content in the workplace, %;

- σ_c is the average standard deviation of metal content in the workplace (the unit), %.

The following values in the annual production program are known:

- Q_m is the planned monthly ore extraction, t;

- c_m is the planned monthly average metal content, %.

We assume the permissible deviation in the performance of planned indicators for the mass ±5%, and for metal content: ±0.05%.

It is assumed that for all blocks the site has the necessary quantity of reserves ready to be stripped (category 111).

The problem is solved in two stages.

In the first stage, the extraction is distributed among n workplaces in order that the following conditions be fulfilled

$$0.975Q_m \leq \sum_{i=1}^{n} q_i \leq 1.025Q_m, \qquad \text{(IV.4)}$$

$$c_m - 0.025 \leq \frac{\sum_{i=1}^{n} c_i q_i}{\sum_{i=1}^{n} q_i} \leq c_m + 0.025, \qquad \text{(IV.5)}$$

where q_i is the planned extraction from the workplace i;

c_i is the respective metal content, %.

Of course, the limits of inequalities (IV.4) and (IV.5) can also be set with other numerical coefficients.

For the sake of solution's univocity some assessment criterion must be adopted. It would be appropriate is such criterion is based on relations (IV.1) and (IV.2). In the first case the criterion is the minimum deviation of projected content of the ore extracted from the planned one.

$$\left| \frac{\sum_{i=1}^{n} c_i q_i}{\sum_{i=1}^{n} q_i} - c_m \right| \rightarrow min. \tag{IV.6}$$

Where standard deviation matters the criterion could be:

- the minimum value of content's dispersion in-between the faces

$$\sum_{i=1}^{n} p_i (c_i - \bar{c})^2 \rightarrow min, \tag{IV.7}$$

where $p_i = \frac{q_i}{\sum_{i=1}^{n} q_i}$;

$\sum_{i=1}^{n} p_i = 1$;

- the minimum value of the sum of modules of differences

$$\sum_{i=1}^{n} p_i |c_i - \bar{c}| \rightarrow min. \tag{IV.8}$$

Other optimization criteria, which affect also the quantity of extraction from separate blocks, are possible as well.

The described formula apparatus has been applied for monthly planning of extraction by using the input data stated in Table IV-1.

A uniform distribution of extraction from workplaces and a normal distribution for metal contents have been assumed.

The planed indicators are:

- monthly extraction – 15 thousand ± 600 t;

- average metal content – 1.15 ± 0.15%.

When solving the problem 100 thousand combinations were generated of which 23.8 thousand are successful as per the restrictions (IV.4) and (IV.5).

It can be seen from Figure IV-3 that in 63% of cases the deviation will be 0.09, and in 35% of cases it will be higher than 0.12.

With the optimization criterion (IV.6), the planned monthly indicators are:

- extraction – 14 729 t;

Table IV-1. Model input data for the analysis

Workplace number	Possible monthly extraction, t		Metal content, %		Projected extraction, t	Projected extraction, t
	Average	Deviation	Average	Deviation		
1	2	3	4	5	6	7
1	2 010	350	1.26	0.09	1 975	2 303
2	840	180	1.52	0.28	785	665
3	1 440	280	1.17	0.09	1 467	1 263
4	1 050	210	1.16	0.14	989	1 065
5	1 500	260	1.07	0.12	1 682	1 306
6	1 360	240	1.20	0.13	1 226	1 530
7	2 490	500	1.36	0.22	2 259	2 389
8	990	120	1.53	0.16	1 078	1 045
9	1 230	440	0.70	0.01	1 087	1 296
10	870	120	1.30	0.07	913	773
11	1 320	320	1.10	0.08	1 286	1 207
				Total	14 729	14 843

Figure IV-3. Deviation of scheduled content

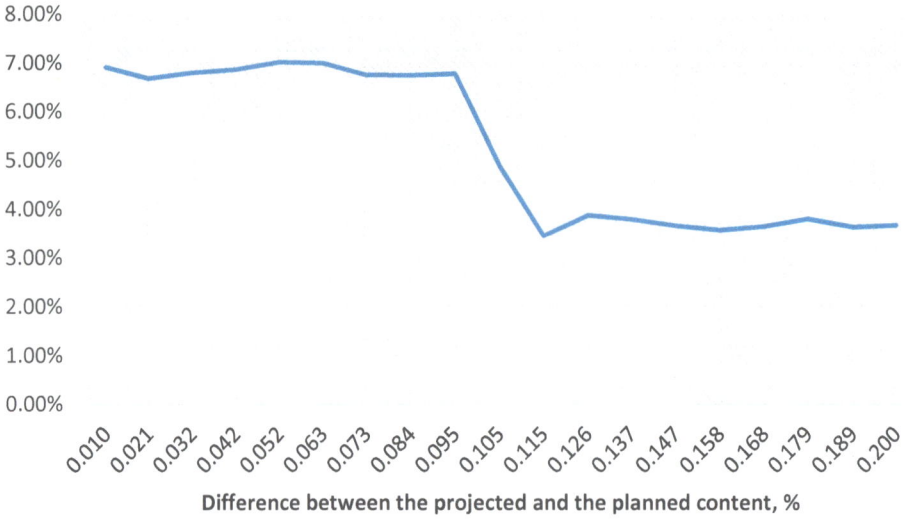

Figure IV-4. Relative frequencies of differences between project and planned content

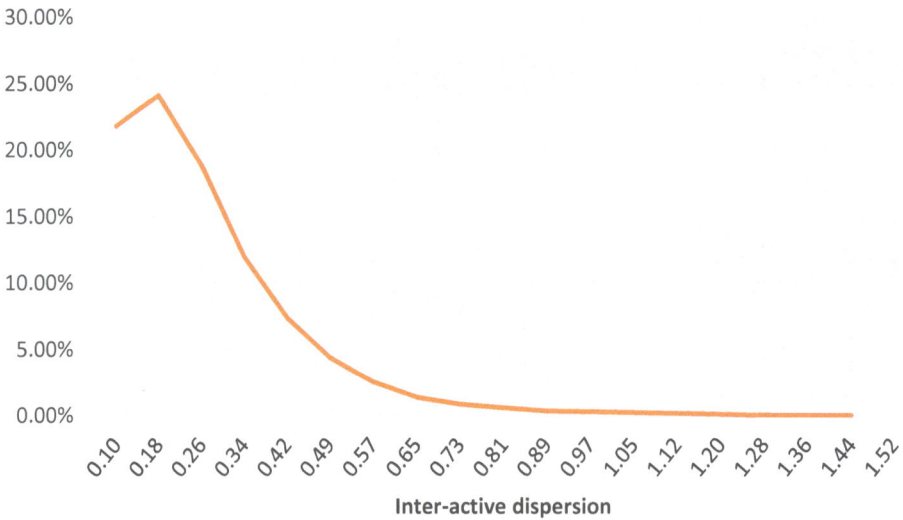

Figure IV-5. Relative frequencies of dispersions in-between the faces

- average metal content – 1.1499 % (!).

The distribution by workplaces is given in Table IV-1, col. 6.

The second stage is a checkup of the sensitivity of the obtained solution. The same formula apparatus is used, this time with the extraction from workplaces obtained at the first run, by taking into account the inevitable deviations in the run of the planned task. Thus, an assessment of solution's stability is obtained depending on the natural changeability of metal content in the earth's womb. An indicator for that is the deviation of projected content compared to the planned one (IV.6).

Figure IV-4 shows the relative frequencies of the differences between the projected and the planned contents.

With the optimization criteria (IV.7), the planned monthly indicators are:

- extraction: 14,843 t;

- average metal content: 1.1035 %.

The distribution by workplaces is given in Table IV-1, col. 7.

Figure IV-5 shows the relative frequencies of dispersion in-between the faces.

The described model can be expended and enriched by more than one useful component, by adding an optimization criterion, for example a minimum number of workplaces, liming loss and depletion, use of a more detailed geological model.

IV.3. DAILY EXTRACTION PLANNING IN UNDERGROUND MINES

N. Valkanov, S. Topalov, K. Boev (2014) proposed an algorithm for distribution of monthly extraction by days within 1 to 2 decades (10 to 20 days). In essence, this is a detailization of the solution for development of extraction works in the mine for a period of 1 month (see "*Monthly planning*") and in such case the main input information is the relation between the mass of ore extracted and the metal content known as the reserve/content relation, which can be found with sufficient reliability by means of a numerical geological model.

Table IV-1 shows this relation in the consecutive exhaustion of an extraction block in a vertical vein of ore. Reserves are grouped in 6 work fields to be consecutively stripped. It is assumed that the mass of industrial reserves of category 1.1.1. is greater than the planned extraction for the period.

The input parameters include:

- total planned extraction: $Q \pm dQ$, t;

- planned metal content: $c \pm dc$, %;

- number of periods n (research shows that a sufficiently reliable solution is obtained in 10 to 20 periods);

- planned extraction for 1 period: $q \pm dq$, t;

- coefficient of non-uniformity of daily extraction taking into account the scheduled downtime: $K_{n,i}$.

Table IV-2. Plan for daily extraction

Workplace number	Block	Ore, t	Metal content, %		Stripping sequence
			Average	Deviation	
1	2	3	4	5	6
101	1	1200	2.5	0.6	1
101	2	850	2.4	0.8	2
101	3	700	2.9	0.6	3
101	4	1500	3.5	0.8	4
101	5	1800	3.0	0.9	5
101	6	1000	1.8	0.5	6

The problem is solved subject to the following limitations:

- for quantity (mass)

$$Q - dQ \leq \sum q_i \leq Q + dQ, \text{t,} \qquad (IV.9)$$

- for quality

$$c - cQ \leq \sum p_i c_i \leq c + dc, \text{%.} \qquad (IV.10)$$

where $p_i = \dfrac{q_i}{\sum q_i}$;

- for available reserves of category (111)

$$\sum_{i=1}^{n_{bl}} z_i > Q + dQ, \qquad \text{(IV.11)}$$

where n_{bl} is the number of blocks.

As we have said, the Monte Carlo method enables the optimization tasks to be solved by various criteria (target functions). In the example, the criterion can be the minimum deviation of the mass of daily extraction or the minimum deviation of the contents of daily extraction, which is attained at minimum dispersion of daily extraction –

$$\sigma_q^2 \rightarrow \min \qquad \text{(IV.12)}$$

or of content for the day –

$$\sigma_c^2 \rightarrow \min. \qquad \text{(IV.13)}$$

Other target functions are also possible, e.g. the minimum sum of the modules of deviations (uniform approximation)

$$\sum |\delta_i| \rightarrow \min. \qquad \text{(IV.14)}$$

Table IV-3. Results of computations

Day	Minimum dispersion of content		Minimum dispersion of daily extraction	
	Planned extraction, t	Content, %	Planned extraction, t	Content, %
1	515	2.31	494	2.83
2	489	2.31	485	2.83
3	473	2.33	497	2.88
4	544	2.33	498	2.88
5	525	2.27	495	3.89
6	459	2.27	500	3.09
7	490	2.27	496	3.09
8	478	2.27	496	3.09
9	505	2.37	498	1.74
10	547	2.37	504	1.74

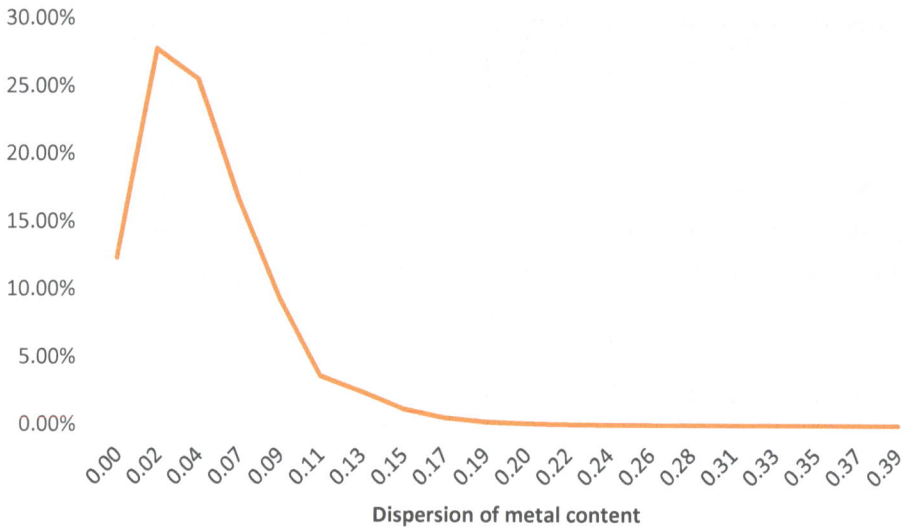

Figure IV-6. Relative frequency of dispersion of metal content

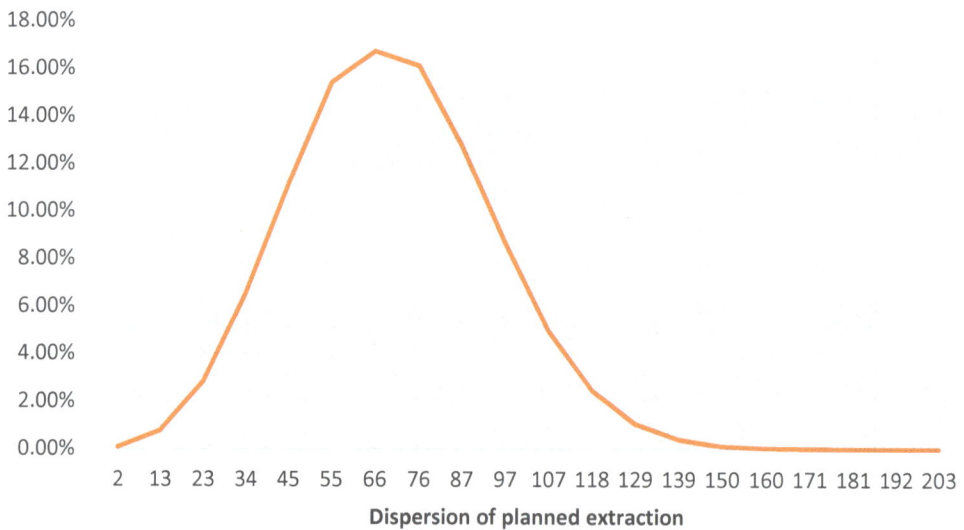

Figure IV-7. Relative frequency of dispersion of planned extraction

For the example under consideration we assume that:

the planned extraction is uniformly distributed in the interval $[Q - dQ, Q + dQ]$;

- the metal content of the block is normally distributed, with an average value of c_i and dispersion σ_i^2;

- the coefficient of non-uniformity is 1.

It is not difficult to extend and enrich the model by other indicators and relations, which take into account the complex nature of distribution and the relations among indicators.

The results from computations are shown in Table IV-3 and on Figure IV-6 and Figure IV-7.

Out of 1 million variants about 760 thousand meet the conditions (IV.9) and (IV.10). In case of target function (IV.12) the distribution is with slightly manifested left asymmetry. The planned daily extraction is within narrow limits: between 485 and 504 t, and the content varies from 1.74 to 3.89%. In case of target function (IV.13) the distribution is manifested as left asymmetrical one. The content varies within very narrow limits: between 2.27 and 2.37%, and the planned extraction is in the range between 459 and 547 t.

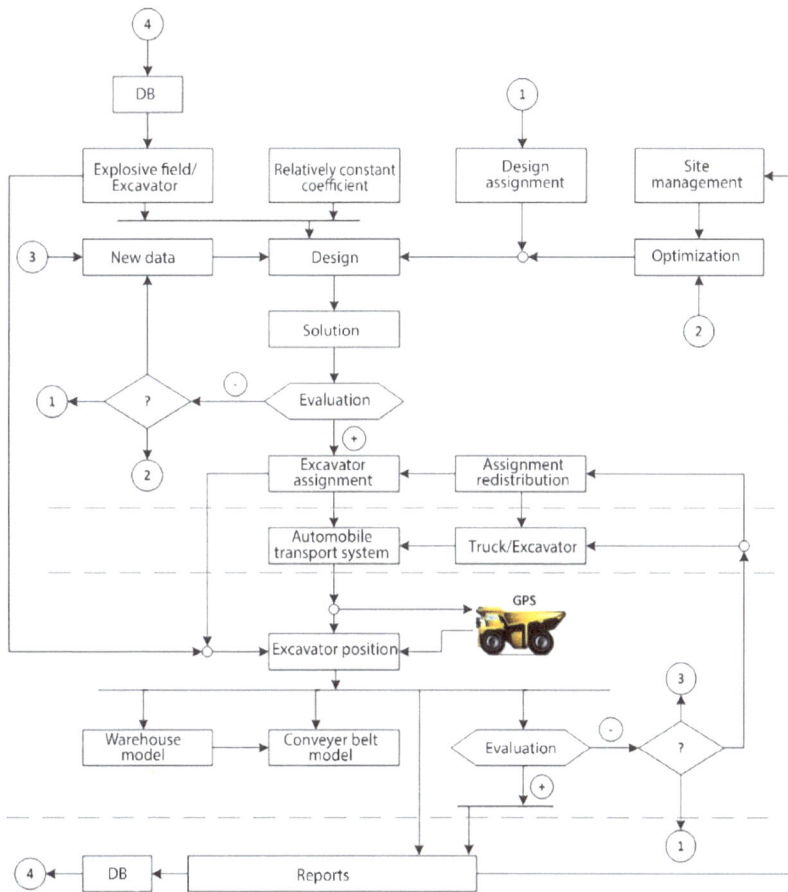

Figure IV-8.

The following two conclusions can be drawn from these results.

1. The selection of a target function has great importance for extraction planning.

2. It is possible to achieve a balance of the requirements of functions (IV.12) and (IV.13) by means of appropriate combinations and development of the method.

IV.4. SHIFT EXTRACTION PLANNING OF NON-FERROUS METAL ORE IN OPEN-PIT

Planning of shift extraction in open-pit mines for non-ferrous metal ores is a non-linear optimization problem that can also be solved by the apparatus of stochastic modeling.

We divide the input data into two groups.

1. Planned indicators for the mine:

– minimum and maximum useful component content: c_{min}, %, and c_{max}, %;

– minimum and maximum mass of ores and minerals: Q_{min}, t, and Q_{max}, t;

– number of extraction excavators: n.

2. Planned indicators for each extraction excavator:

– minimum and maximum shift productivity q_{min}, t, and q_{max}, t;

– relation between ore extracted and its metal content:

$$c = f(q), \%. \qquad (IV.15)$$

The Monte Carlo method makes it possible to find a solution in different target functions and limitation conditions. The example uses minimization of dispersion in-between the faces of the content of ore extracted subject to limitations reflecting the traditional requirements for quantity and quality of extraction.

The target function is

$$\sigma_c^2 \to min. \tag{IV.16}$$

The limitation condition are:

- for the content of total extraction:

$$c_{min} \le c_{pl} \le c_{max}; \tag{IV.17}$$

- for the mass of total extraction:

$$Q_{min} \le Q_{pl} \le Q_{max}; \tag{IV.18}$$

- for the productivity of each excavator:

$$q_{min} \le q \le q_{max}. \tag{IV.19}$$

In formulas (IV.16)-(IV.19):

$$Q_{pl} = \sum q_i, t;$$

$$c_{pl} = \sum p_i c_i, \%;$$

$$\sigma_c^2 = \sum p_i \left(c_{pl} - c_i \right)^2;$$

$$p_i = {q_i}/{Q_{pl}};$$

$$i = 1, 2, \dots, n.$$

The modeled independent variable is the shift productivity of the ith extraction excavator −

$$q^* = q_{min} + (q_{max} - q_{min})s, t, \tag{IV.20}$$

where s is a random number in the interval $[0;1)$.

The useful component content for the realization is found by the formula (IV.15):

$$c^* = f(q^*), \%. \tag{IV.21}$$

The preliminary assessment of the number of simulations can only be approximate because it depends on the amount of the allowance in limitations (IV.17) and (IV.19). Thus, it is purposeful to include "counters" of the variants, which meet the conditions (IV.17) -(IV.19) into the programs. If this number is insufficient the total number of simulations must be increased.

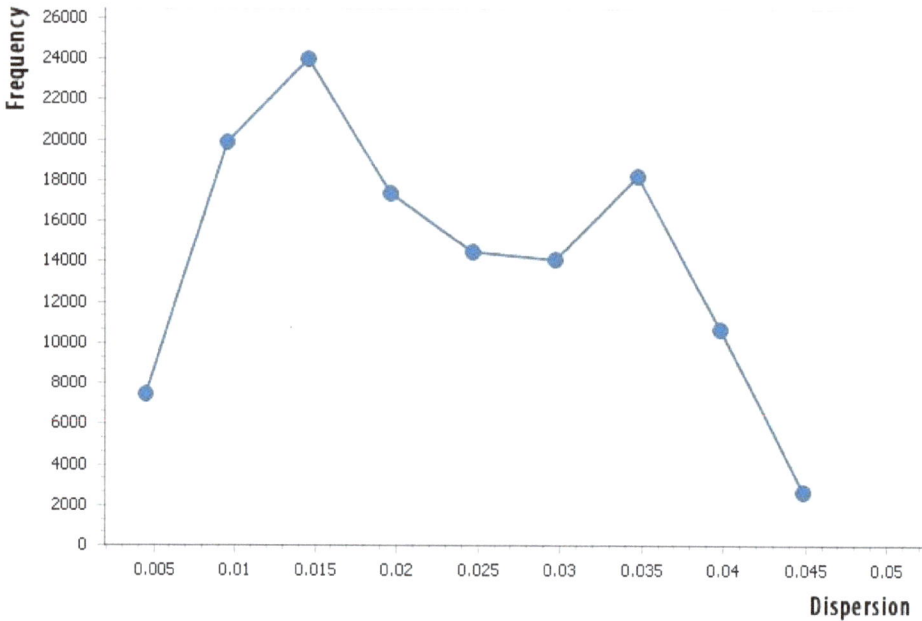

Figure IV-9. Frequency of change of dispersion of useful component content

On the basis of the described approach we have performed an example with the following input data:

– the number of excavators is: $n = 5$;

– the productivity per individual excavator: $q = 1\,500 \div 2\,100$ t, a multiple of 10 t;

– the total planned productivity: $Q = 8\,000 \div 12\,000$ t;

– the useful component content in the field of the extraction excavator: $c = 0.25 \div 0.50$ %;

– the permissible change of planned useful component content: $c = 0.32 \div 0.39$ %.

After 200 thousand runs, of which about 130 thousand satisfy the set limitation conditions, we get the distribution of the dispersion given in Figure IV-9. The shift productivity of individual excavators is 1 750, 2 010, 1 980, 1 650, 1 640 t, a total of 9 030 t; the average useful component content is 0.387 %.

REFERENCES

[1] Andonov, I. (2003). Application of Geochemical Field Theory in Determining the Effect of Standard Deviation and the Content of the Rated Component in the Ore on the Rate of Extraction. Xth National Mine Surveying Conference with International Participation, Varna, Bulgaria [in Bulgarian].

[2] Boev, K., I. Vaklev (1978). "Method for Planning of Quality Indicators for Mining in Underground Mines." Rudodobiv(1): 23-27 [in Bulgarian].

[3] Mazhdrakov, M., D. Benov (2006). Automated system for management of quality of the extracted ore in opencast mine for mass detonation. 6th International Scientific Conference - SGEM2006, Albena, Bulgaria.

[4] Mazhdrakov, M., D. Benov, G. Trapov (2013). CADMin. Automated Planning of Development of Mining Works in Open Pits. Sofia, MGU St. Ivan Rilski [in Bulgarian].

[5] Smirnov, N., I. Dunin-Barkovskii (1962). Course of probability theory and mathematical statistics. Moscow, Nauka [in Russian].

[6] Valkanov, N., S. Topalov, K. Boev (2014). Method of Planning Indicators of Probabilistic Nature. Fourth National Scientific and Technical Conference with International Participation "Technologies and Practices in Underground Mining and Mine Construction", Devin, Bulgaria [in Bulgarian].

SECTION V.
STABILITY OF EMBANKMENTS

St. Hristov, G. Trapov, M. Mazdrakov, D. Benov,
N. Valkanov, I. Vasilev, St. Bosnev

C ontemporary engineering methods for computing the slope stability of open-pit mines are based on the theory of limit equilibrium. From that theory it follows that for as long as landslide forces do not exceed the resistance forces the slope will be in a state of equilibrium. But if landslide forces exceed even slightly the landslide ones, this is sufficient for a clearly manifested deformation process to begin.

A criterion for assessment of slope stability is the coefficient of stability. It is determined by the ratio of resistance forces $\sum S_i$ to landslide forces $\sum T_i$, active in the most adverse sliding surface.

$$\eta = \frac{\sum S_i}{\sum T_i},$$

(V.1)

If $\eta = 1$ the slope is in a limit equilibrium, if $\eta > 1$ it is stable, and if $\eta < 1$ it is instable.

Currently, a deterministic approach is used when solving the problems of slope stability. All output data such as the angle of internal friction φ, cohesion C, volumetric density γ, compression strength σ_{cs}, tensile strength σ_o , are assumed as determined parameters for a long period of time. Usually the arithmetic mean values of strength indicators are used in computations.

In the process of repeated laboratory and field determination of strength indicators of rocks or of other parameters, such as cracks in the massif, thickness (width) of lithological varieties, situation on the sliding surface, relief of the area, characterizing the slopes of the step, the wall and the embankment, a massif of numerical values of certain law of distribution is obtained. In the dynamics of development of mining works the resistance and landslide forces will also be changed. Therefore, the results of computations for the stability coefficient of the slopes must also represent a massif of numerical values with a corresponding law of distribution. In computing one works with a great number of factors whose nature of change is probabilistic, i.e. there is a considerable element of indeterminateness and risk.

The probabilistic nature of output data for determining the slope stability necessitates these deterministic solutions to be replaced by probabilistic ones.

V.1. INDICATORS IN DETERMINING EMBANKMENT STABILITY

When checking the embankment stability it is necessary to defined and substantiate computing indicators.

To compute the stability coefficients of the bulk body the geotechnical indicators of angle of internal friction $\varphi,°$, cohesion $c, kN/m^2$, and volumetric density $\gamma, kN/m^3$ are used. A probabilistic determination of geotechnical indicators under the Monte Carlo method has been used subject to the premise of normal distribution of separate physico-mechanical indicators, which is known in geotechnics. When determining the limits, or the so-called interval of possible values of individual indicators (mainly for the angle of internal friction and cohesion), the technology of embankment formation is also taken into consideration. Following that technology the primary geotechnical indicators are determined in depths, which approximately limit individual layers (steps) of the embankment, in this case: 16.5 m, 34.1 m, 45.0 m, ... (S. Bosnev, S. Hristov, N. Mihaylov et al., 2010).

For each of these depths a probabilistic assessment has been made and the most probable value of the strength of cutting the embankment material has been found as a function of the geological load according to the well-known law of Mohr – Coulomb (S. Hristov, 1978).

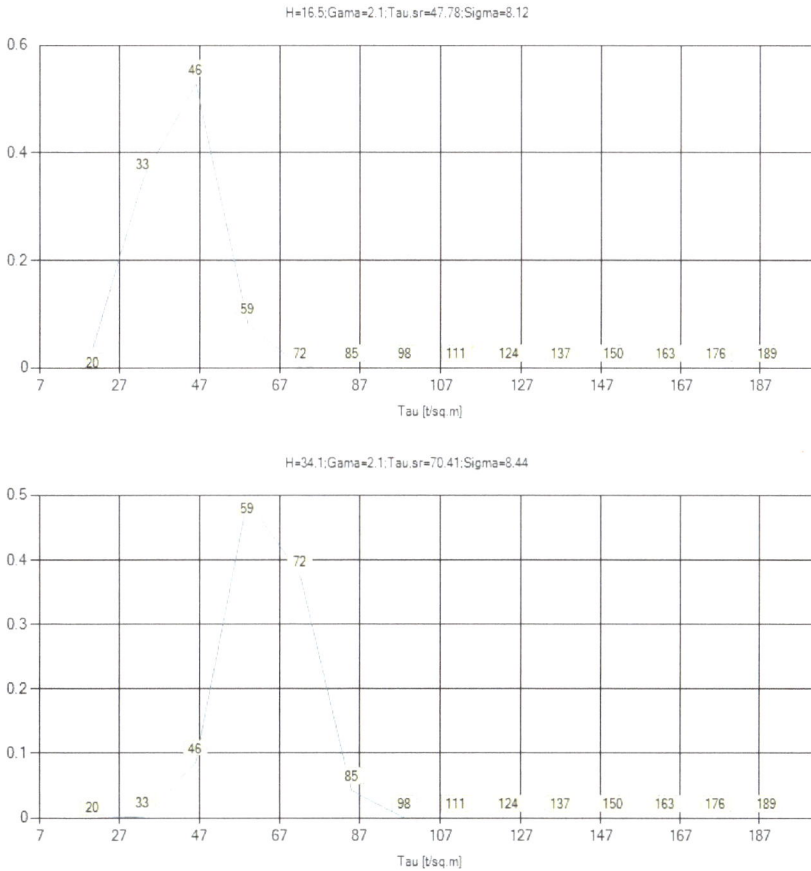

Figure V-1. Geotechnical indicators at depth of 16.5 m (above) and 34.1 m (below)

Figure V-2 presents the computed average values of the cutting strength (τ, kPa), as well as the standard deviation (σ) in the set depth and volumetric density of earth masses or the obtained geological load ($\sigma = \gamma h_i$, kPa).

The function (Figure V-2) has been constructed according to the obtained most probable cutting strength values.

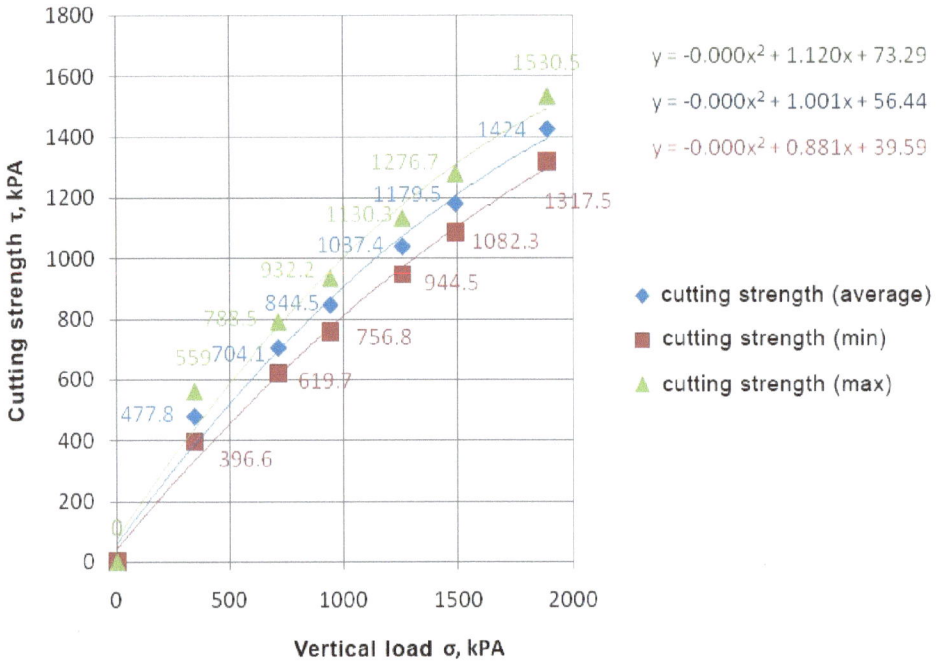

Figure V-2. Graph of the function (V.2)

$$\tau = f[\sigma(h)].\qquad\qquad(V.2)$$

On Figure V-2 the values of the angle of internal friction and the cohesion of every modeled layer of the embankment can be read. The graph reflects the non-linear change in the angle of internal friction and the cohesion in great depths.

V.2. STABILITY ASSESSMENT IN EMBANKMENT CONSTRUCTION

During the construction of mining and building sites at Maritsa East's[18] Pliocene basin large non-productive depots (embankments) were made of clay. The clays from the overburden in open-pit mines

---

18 The largest energy complex in South Eastern Europe, located in Bulgaria

have different physical properties and low resistance. Considerable deformations are observed both in embankments' slopes and base.

The specific characteristics of embankments' existence lead to technological and economic risk (M. Mazhdrakov, D. Benov, G. Trapov, 2013).

Long-term observations make it possible to establish the mechanism of deformation by means of a complex potential sliding surface (Figure V-3).

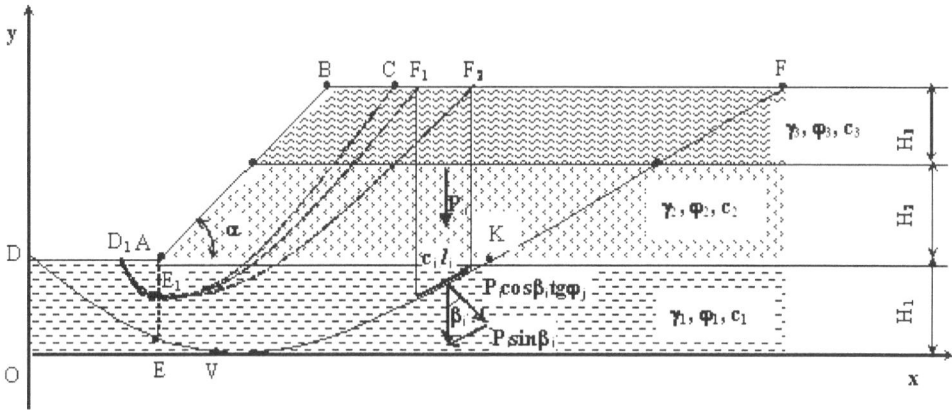

Figure V-3. The parameters of the slope (the slope's angle (the angle α, heights H_1 and H_2) and the configuration of potential sliding surface

We will examine in detail the following scheme. The embankment is made of two layers of Pliocene clays: lower layer of gray clays with thickness H_2; upper layer of organic clays with thickness H_3, which change their properties in time: from the time of piling until they reach relative consolidation. Values of the angle of the slope, the heights and the properties of each layers are imitated in modeling.

The stability coefficient (F) is determined as the ratio of the holdback (passive) and landslide (active) forces.

The computed minimum and average values of stability coefficient F_{min} and $\bar{F}$ as well as the standard deviation of the average value are shown in Table V-1.

The risk of occurrence of deformations and landslides for each stated studied is defined by using the classic probability formula

$$R = \left(1 - \frac{m}{n}\right) 100, \%, \tag{V.3}$$

where m is the number of tests where the passive forces are greater than or equal to the active ones, i.e. $F \geq 1$;

n is the total number of computer experiments conducted.

Table V-1. Computed values of stability coefficient and standard deviation

State	$\overline{F}$	F_{min}	σ
Non-consolidated	0.96	0.94	0.035
Partially consolidated	0.93	0.91	0.032
Consolidated	0.91	0.88	0.030

In the scheme being examined the risk in the states specified in Table V-1 is $R = 25\%, R = 54\%$ and $R = 81\%$, respectively.

The suggested method can be successfully used for different cases of risk assessment for a multi-layer massif.

V.3. MODELING STABILITY IN MULTI-LAYER EMBANKMENTS AND MASSIFS

The overlying Pliocene complex of the walls in Maritsa East's open-pit mines is a multi-layer environment of different physical and strength properties for each stratum. When building the embankments a multi-layer macrostructure is made simultaneously with the stripping of overburden.

Wall stability in the multi-layer massif depends on the thickness of lithological varieties, on their physico-mechanical characteristics and on the technological parameters of steps and the wall as a whole. Based on the numerous laboratory studies of physical and strength properties of all lithological varieties of the Pliocene profile in the massif the respective interval statistical distributions have been obtained for the varieties:

    – in the wall: gray clays, black clays, coal, interbed, underlying clays;

    – in the embankment: gray clays, black clays, contact.

By using stochastic modeling certain values of volumetric density, the angle of internal friction and cohesion are randomly selected for each lithological variety.

The technological parameters of the steps are obtained from a real open-pit mine section. As in certain time interval these parameters change in a small range they are examined as determined values.

Table V-2. Structure of a horizon from the embankment

Scheme A	Thickness [m]	Scheme B	Thickness [m]
Gray clays	18 to 22	Gray clays	3,5 to 4.5
Black clays	3.5 to 4.5	Black clays	18 to 22

Table V-3. Average values and confidence intervals for the wall stability coefficient.

Number of tests	Average value of F	95% confidence interval for F	
		left end	right end
20K	1.09	0.97	1.24
50K	1.12	1.01	1.23
100K	1.11	1.01	1.22

Table V-4. Average values and confidence intervals for the embankment stability coefficient.

Scheme	Number of tests	Average value of F	95% confidence interval for F	
			left end	right end
A	20K	0.92	0.77	1.07
A	50K	0.91	0.77	1.06
A	100K	0.91	0.78	1.04
B	20K	1.14	1.00	1.28
B	50K	1.13	1.02	1.25
B	100K	1.13	1.03	1.23

In practice, there are two possible schemes of embankment construction. In each of them there are three horizons. Horizon structure examined from higher elevations towards the lower ones in each scheme is given in Table V-2.

The adopted computing scheme of deformation, i.e. a prism of active earth pressure, central block, prism of passive pressure, weak contacts or mirror surfaces, has been used for the wall stability assessment at Maritsa East (S. Hristov, 1978).

Stochastic modeling is realized in the following sequence.

1. Values of step's height, slope angle and platform width for each step are imitated.

2. For each variety the thickness and the values of indicators γ_i, φ_i, c_i. are obtained.

3. The stability coefficient is computed (F).

The results from the application of the described manner of computing the stability for the working wall and for the embankment are shown in Table V-3 and Table V-4.

The analysis of results in Table V-3 shows that in the section being examined the wall stability coefficient falls within the range [1.01; 1.22] with 95-percent certainty.

The results in Table V-4 show that only scheme B ensures a confidence interval for the stability coefficient with ranges above 1. therefore, if there are no technological obstacles this scheme must be adopted upon embankment formation.

V.4. DETERMINING THE STABILITY COEFFICIENT OF EMBANKMENTS BY MEANS OF SLOPE/W [19]

The purpose of this illustartive example is to show how you can do a probabilistic analysis and a sensitivity analysis of the stability of a slope. Features of this simulation include:

- analysis method: Morgenstern-Price with Half-sine function;
- statistical variability in some parameters;
- multiple soil layers with different soil models;
- pore water pressure with piezometric line;
- entry and exit slip surface with projection angle;

19 Third party copyright © GEO-SLOPE International Ltd., Calgary, Alberta, Canada, www.geo-slope.com. The material is published with explicit written permission.

- auto search for tension crack zone.

Example problem definition and material properties. Figure V-4 shows the geometry, the piezometric line and the entry and exit search zones. The red arrows on the entry zone indicate the projection angle of the slip surfaces.

The embankment is modeled with a Mohr-Coulomb soil model. The unit weight is 20 kN/m³, c' is 30 kPa and $\varphi = 40°$. The foundation material is modeled with a $S = f(overburder)$ soil model. The unit weight is 20 kN/m³, the Tau/Sigma Ratio is assumed to be 0.45.

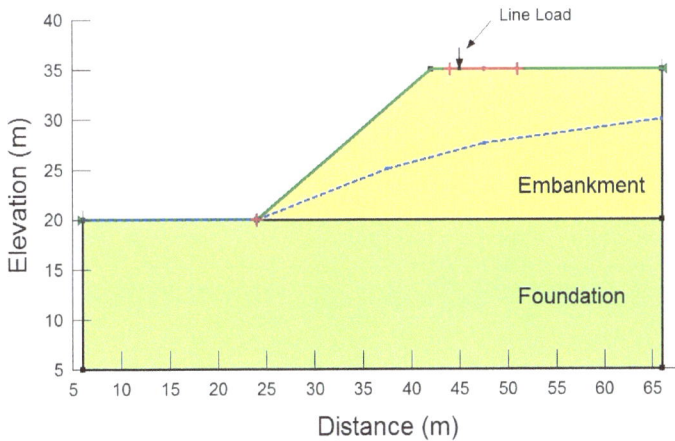

Figure V-4. Geometry and piezometric line of the example

Case 1 – Probabilistic analysis

Figure V-5 shows the most critical slip surface and the computed factor of safety. The deterministic critical factor of safety is 1.121. Note the tension crack developed bear the crest when the base angle reaches 75 degrees.

A probalistic analysis can be performed quite easily with SLOPE/W when the variability of the soil properties or other input parameters is known. Although SLOPE/W allows various type of variability distribution, the most common approach is to assume a normal distribution with a specified mean value and standart deviation of a parameter. In this example, the mean value and the standart deviation for phi, cohesion, unit weight, pore water pressure; line load and seismic coefficient are specified as shown in Figure V-6.

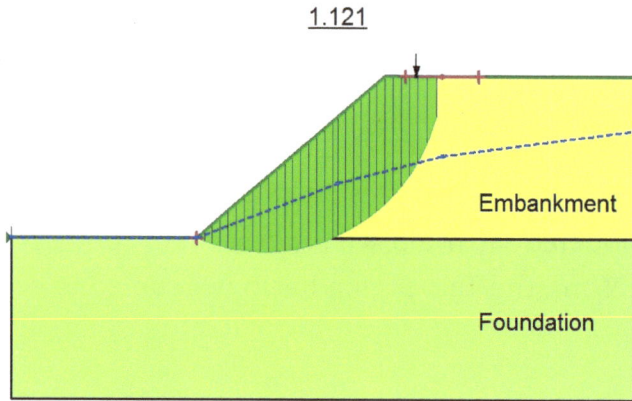

Figure V-5. Factor of safety of the example probabilistic analysis

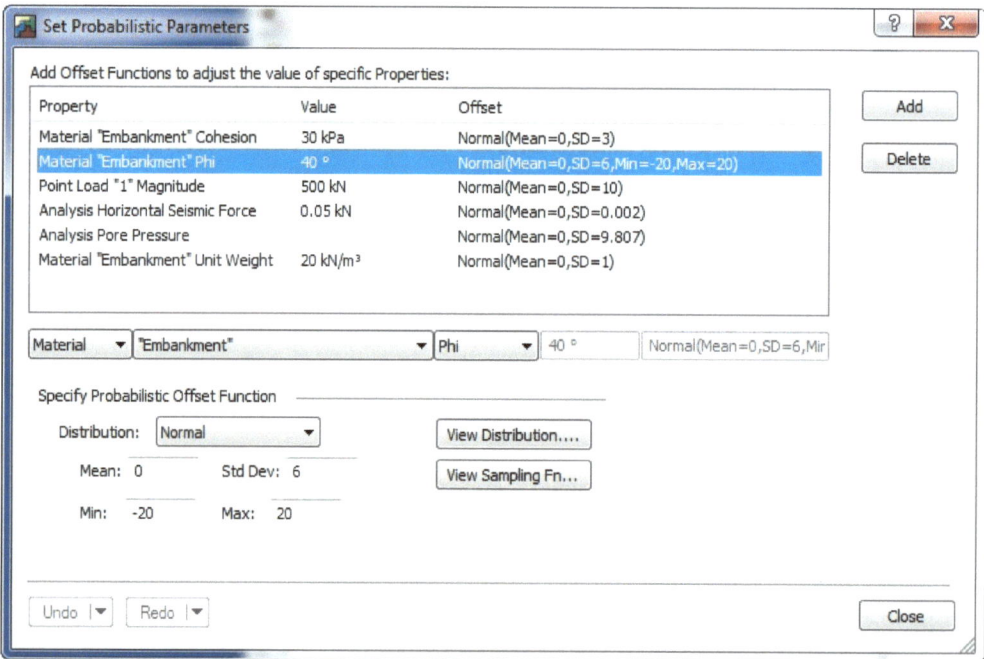

Figure V-6. Defining and setting of parameters for probabilistic analysis

For example, for the Embankment Phi parameter, the mean value and the standart deviation are specified to be 40° and 6° respectively. Also, the minimum and maximum offsets are also capped to from −20° and 20°. In other words, the embankment phi parameter will distributed normally between 20° to 60°. Figure V-7 shows the distribution function and the corresponding sampling function for the Embankment Phi during the Monte Carlo similation.

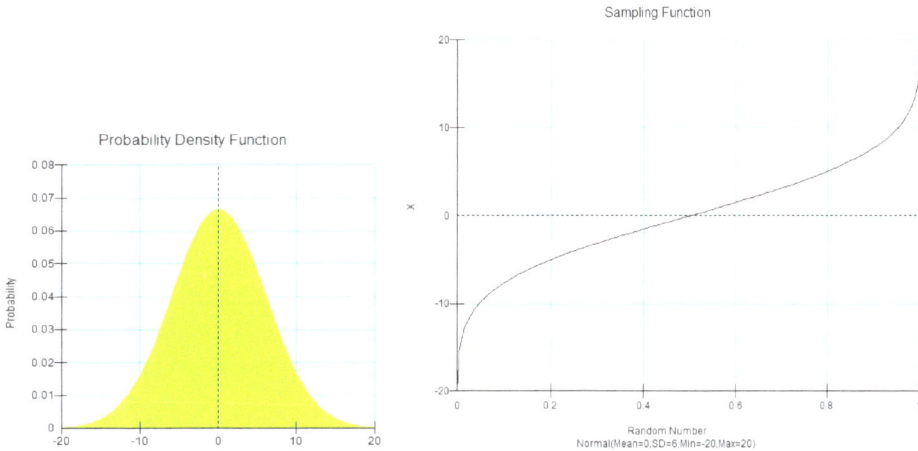

Figure V-7. Normal distribution function and sampling function of Embankment Phi for random number generation from 0 to 1

After 5 000 Monte Carlo simulation, the following results are obtained (Figure V-8):

- mean F of S = 1.1219;
- reability index = 1.167;
- probability of failure = 12.6%;
- standart deviation of F of S = 0.10442;
- min F of S = 0.76793;
- max F of S = 1.4869.

Using the "Draw Probability" feature in Result View, you can get a probability density function of the 2000 computed factors of safety. Since the variability of the soil properties is normally distributed, the probability density function of the factor of safety (Figure V-8) is also normally distributed as expected. Figure shows the probability distribution function of the Monte Carlo factors of safety. Note that the corresponding probability for factor of safety smaller than 1 represents the probability of failure. The probability of failure is computed to be 12.6%.

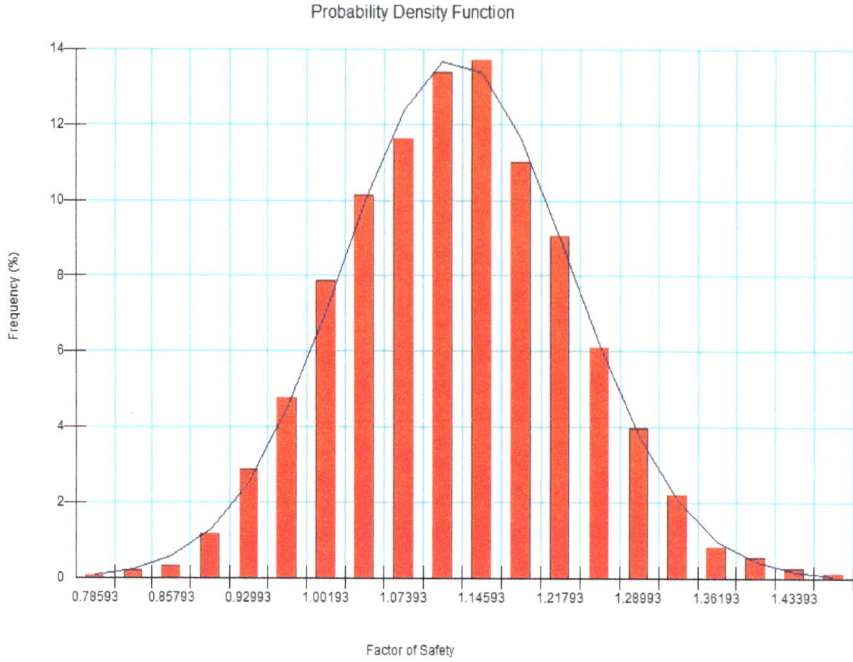

Figure V-8. Probability density function of the 5000 Monte Carlo factors of safety

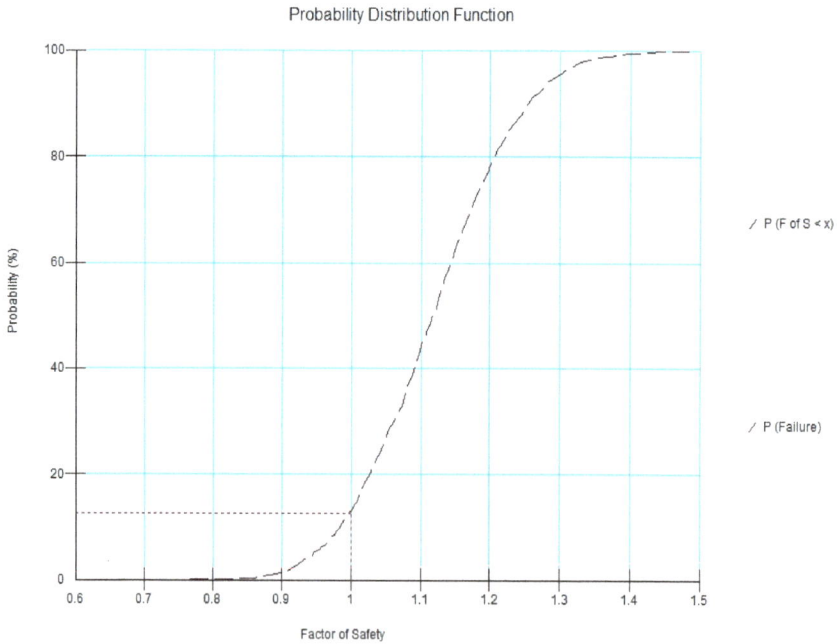

Figure V-9. Probability distribution function of the 5000 Monte Carlo factors of safety showing probability of failure

In this case, for each Monte Carlo trial, the soil properties of the entire soil layer are only sampled once, the sampling value is applied to all slices within the same soil layer, the probability of failure is 12.6%. The other extreme is to sample new soil properties for every slice in each Monte Carlo trial. The probability of failure is now reduced to 3.2%. These two options are useful to compute the range of the probability of failure. The more realistic probability of failure is likely in between of 3.2% and 12.6%. When more field investigation and soil testing are available, and when more statistical data become available to justify the use of a certain sampling distance, the probability of failure can be more accurately estimated. For example, if you assume a sampling distance of 10 m, you will get a probability of failure of 9.75%. To evaluate what should be the correct sampling distance in a probabilistic analysis, you will need a variogram analysis (Figure V-8), which is beyond the scope of this discussion.

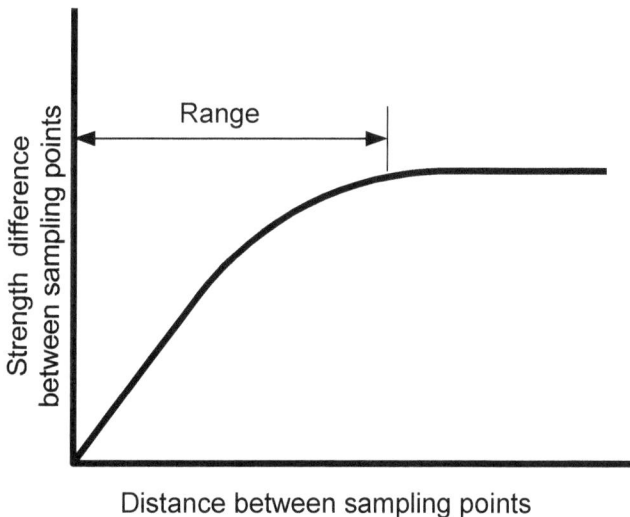

Figure V-10. Probability distribution function of the 5000 Monte Carlo factors of safety showing probability of failure

Unfortunately, sufficient data is seldom available to formally evaluate the spatial variability, and so it comes down to making a judgment. Making an intuitive judgment, however, may be better in some cases than ignoring the possibility of spatial variability completely. H. El-Ramly, N. Morgenstern, D. Cruden (2002) go so far as to say that ignoring spatial variability, "... *can be erroneous and misleading*." This may be overstating the case for some analyses, but it does highlight the importance of giving spatial variability its due consideration in an analysis. The beauty of a tool such as SLOPE/W is that various sampling distances can be examined easily and quickly to assist with making a judgment on an appropriate sampling distance. SLOPE/W perhaps does not accommodate all the nuances and refinements possible in a probabilistic stability analysis. However, the extensive range of features and capabilities available are likely adequate for use in practice, especially considering that probabilistic analyses are yet not routine in practice. It is considered important to not over-complicate the issues when firm practice approaches have not yet been established. Complexity and capability can and will be added, as probabilistic analyses mature more in practice. The big attraction of a tool such as SLOPE/W is the ease with which probabilistic analyses can be performed. Once a problem has been set up for a deterministic analysis, there is very little extra modeling effort involved in doing a probabilistic analysis. No extra tools are required. A variety of probabilistic distribution functions are available with truncation as necessary. Even a general data point distribution function is available for any unusual distribution of properties. The most difficult part of the analysis is obtaining sufficient data to define the dispersion appropriately.

Case 2 – Sensitivity analysis

SLOPE/W allows you to specify a value range of the material parameters, and will compute the factor of safety automatically when each value of the parameters is used. Let's assume that you are interested in the following soil properties ranges:

- unit weight – mean value at 20, study from 17.5 to 22.5 with 10 increments;

- cohesion – mean value at 30, study from 15 to 45 with 10 incretements;

- friction angle – mean value at 40, study from 35 to 45 with 10 increments.

With SLOPE, you can specify the sensitivity study range with a Delta and a Step to both sides (+ and -) of the means value as shown below.

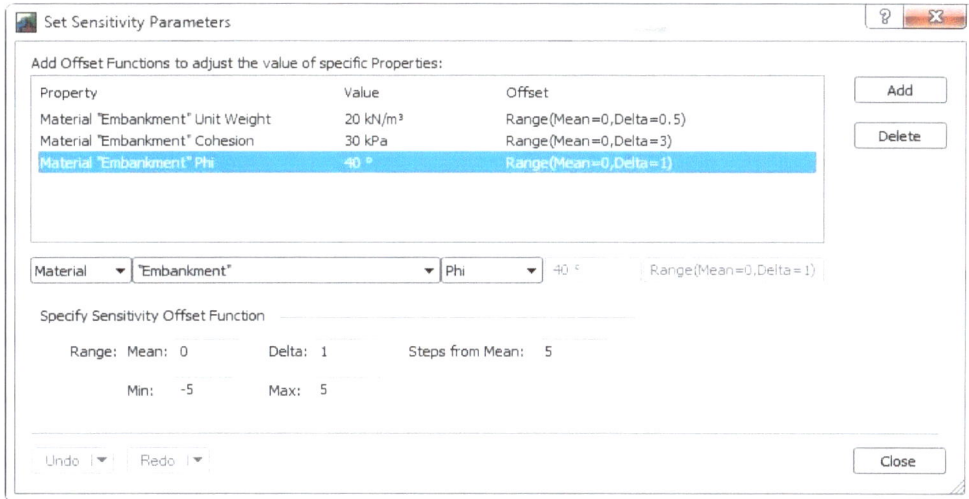

Figure V-11. Defining and setting of parameters for sensitivity analysis

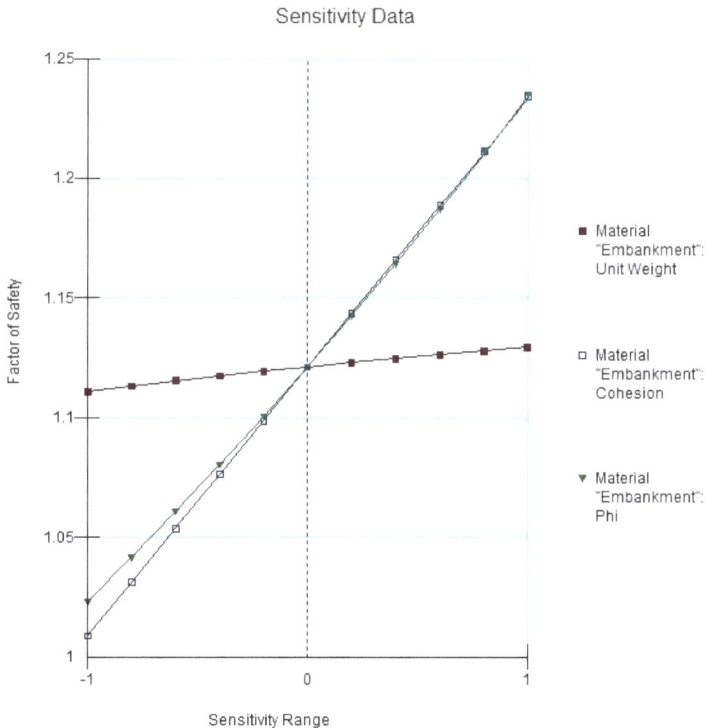

Figure V-12. A plot showing the sensitivity of the computed factor of safety versus the range of the parameters used in the computation

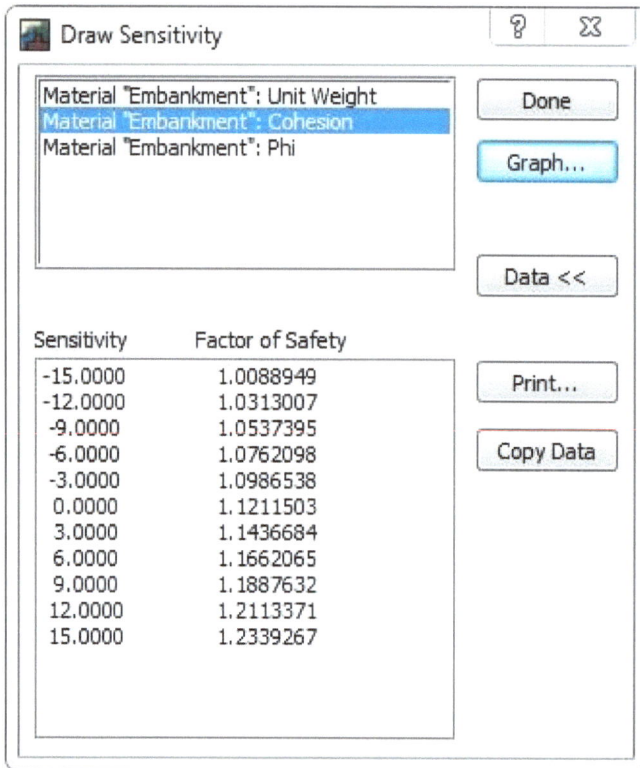

Figure V-13. A data table showing the value of the parameter and the computed factorof safety

When you do a sensitivity study of the above, at the end, SLOPE/W provides a sensitivity graph showing the computed factor of safety at different values of the parameters. For example, you can see that the factor of safety is most sensitive to the cohesion and Phi of the embankment, and is least sensitive to the material unit weight. The crossing point represents the factor of safety (1.121) when the mean value of all the parameters is used. SLOPE/W also provides a table listing the value of a parameter used and the respective factor of safety computed.

REFERENCES

[1] Bosnev, S., S. Hristov, N. Mihaylov, et al. (2010). Project: 1F-05-10. Oxidation deposit stability check due to the application of the new leaching intensification technology. Sofia, BT Engineering Ltd [in Bulgarian].

[2] El-Ramly, H., N. Morgenstern, D. Cruden (2002). "Probabilistic Slope Stability Analysis for Practice." Canadian Geotechnical Journal 39(3): 665-683.

[3] Hristov, S. (1978). Resistance and drainage of slopes. Sofia, Tehnika [in Bulgarian].

[4] Mazhdrakov, M., D. Benov, G. Trapov (2013). CADMin. Automated Planning of Development of Mining Works in Open Pits. Sofia, MGU St. Ivan Rilski [in Bulgarian].

SECTION VI.
SLOPE STABILITY IN OPEN-PIT MINES

M. Mazdrakov, D. Benov, G. Trapov, St. Hristov

VI.1. FINDING THE EXTREME VALUES OF STABILITY COEFFCIENT

One the main tasks in design and operation of open-pit mines is to ensure stability of working, non-working walls, embankments and the adjacent infrastructure (M. Mazhdrakov, D. Benov, N. Mihaylov, 2012). For open-pit mines in Maritsa East this is a crucial matter as work takes place in weak rocks, on larger areas and with considerable changes to the properties in vertical and lateral aspect, i.e. under complex engineering and geological conditions. The situation is further complicated if one takes into consideration the impact of technical-mining and technological-mining factors (P. Zlatanov, M. Mazhdrakov, G. Trapov, 1985).

Many of the well-known methods of determining stability enable the form and position of sliding surface to be found in the massif at any point from which the condition for Coulomb's limit equilibrium is fulfilled. The selected method must enable the weakest sliding surface to be found while observing the static equilibrium.

Under Fisenko's method (S. Hristov, 2001), if the rock massif is somehow divided into n blocks, the stability coefficient is determined by the function

$$F(\gamma,\varphi,c) = \frac{\text{tg}\varphi \sum_{i=1}^{n} P_i \cos\alpha_i + \sum_{i=1}^{n} c_i \ell_i}{\sum_{i=1}^{n} P_i \sin\alpha_i}. \qquad (VI.1)$$

where P_i is the mass of the i-th block, t;

α_i is the angle of inclination of sliding surface in the base of the i-th block;

φ is the angle of massif's internal friction;

c_i is the cohesion in the massif of the i-th block, t/m²;

ℓ_i is the length of sliding line of the i-th block, m.

The analysis of the expression (VI.1) shows that F is a function of density (used in computing P_i), the angle of internal friction and cohesion. The values of these quantities change in certain ranges and it is often assumed that they have uniform distribution in this interval, i.e. at equal lengths the probabilities remain the same. If certain conditions and fixed values of the indicators γ_0, φ_0 and c_0, are met Fisenko's method of computation and other similar methods make it possible to determine the sliding surface's position, i.e. the surface with minimum value $F_{min}(\gamma_0, \varphi_0, c_0)$ of the function (VI.1), whose value is the stability coefficient.

From here the meaning of establishing the minimum $\min F_{min}(\gamma,\varphi,c)$ and the maximum $\max F_{min}(\gamma,\varphi,c)$ of stability coefficient becomes clear, with a possible change to the computing indicators in certain limits. At the same time it is important to find the corresponding sliding surfaces as well. It is difficult to solve such a multi-dimensional problem by using the traditional mathematical means.

The method of random search with reverse step has been used to find the minimum $\min F_{min}(\gamma,\varphi,c)$ and the maximum $\max F_{min}(\gamma,\varphi,c)$ of stability coefficient $F_{min}(\gamma_0, \varphi_0, c_0)$. We will illustrate the solution by determining the position of sliding surface in the overburden step made of blue clay with overburden step's height of 12 m and slope angle of 53° . The values of physical and strength indicators are given in Table VI-1.

Table VI-1. Values of physical and strength indicators

Designation	Minimum value	Maximum value	Average value
$\gamma, t/m^3$	$\gamma_{min} = 1.70$	$\gamma_{max} = 2.00$	$\bar{\gamma} = 1.85$
$\varphi, °$	$\varphi_{min} = 5$	$\varphi_{max} = 8$	$\bar{\varphi} = 6.5$
$c, t/m^2$	$c_{min} = 3$	$c_{max} = 6$	$\bar{c} = 4.5$

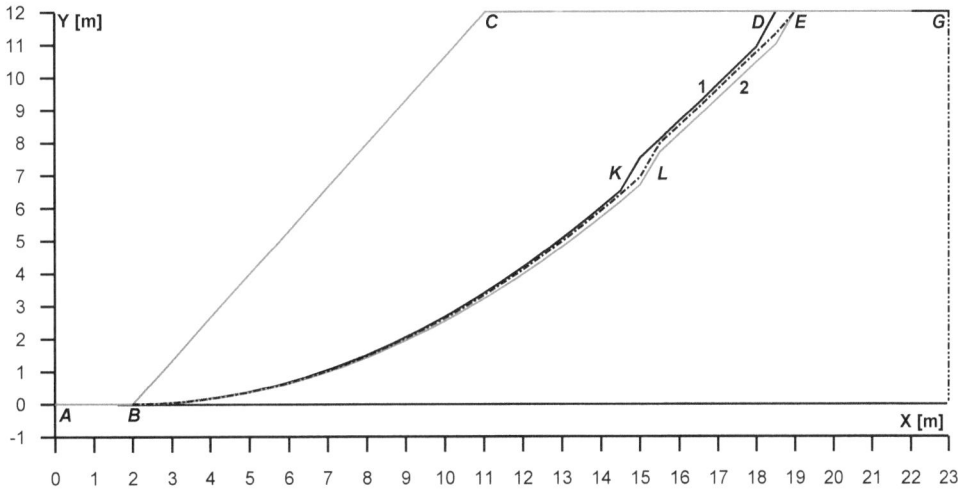

Figure VI-1. Sliding surfaces: BKD where $F_{min}(\gamma, \varphi, c) = 0.86$; BLE with stability coefficient $max F_{min}(\gamma, \varphi, c) = 1.75$; the surface where $F_{min}(\bar{\gamma}, \bar{\varphi}, \bar{c}) = 1.28$ is in dotted line

The results are shown on Figure VI-1. The sliding surface BKD corresponds to a minimum values $min F_{min}(\gamma, \varphi, c) = 0.86$ of stability coefficient obtained at $\gamma = 1.71 \frac{t}{m^3}, \varphi = 5.44°$ and $c = 5.96 \, t/m^2$. The maximum value $max F_{min}(\gamma, \varphi, c) = 1.75$ was found as per the sliding surface BLE, with indicators $\gamma = 1.99 \frac{t}{m^3}, \varphi = 7.06°$ and $c = 3.02 \, t/m^2$.

If the average values of the indicators $\bar{\gamma} = 1.85 \frac{t}{m^3}, \bar{\varphi} = 6.5°$ and $\bar{c} = 4.5 \frac{t}{m^2}$, are used to compute the stability of the step the sliding surface shown as a dotted line between the lines BLE and BKD is obtained. Then the stability coefficient is $F_{min}(\bar{\gamma}, \bar{\varphi}, \bar{c}) = 1.28$, i.e. the step should be in stable condition. But in the range of the limits set the combinations of the values of the properties allow for the step to be in unstable condition in a great number of cases. The sliding surface BKD corresponds to the minimum value $F_{min}(\gamma, \varphi, c) = 0.86$, and *BLE* corresponds to the maximum $F_{\min}(\gamma, \varphi, c) = 1.75$ of stability coefficient. Upon changes to the limits set (Table VI-1) the following inequalities are fulfilled for sure

$$\min F_{\min}(\gamma, \varphi, c) \leq F_{\min}(\gamma, \varphi, c) \leq \max F_{\min}(\gamma, \varphi, c). \quad (VI.2)$$

The example shows that the random search method can be successfully applied in a number of tasks where the extremum of certain target function is sought with numerous arguments under set limitations. Both the target function and the limitations might be non-linear.

VI.2. DETERMINING THE BOUNDARIES OF AN OPEN-PIT MINE IN CASE OF STEEP DEPOSITS

For the output information about designing an open-pit mine is one of probabilistic nature there is also certain risk when making the final decision on its boundaries. When determining the boundaries of the mine and when assessing the reliability of the decision made it is necessary to use probabilistic methods. Therefore, we suggest a method of determining the boundaries of the mine based on probabilistic solutions (S. Hristov, 2013).

The probabilistic nature of the values of rock's strength indicators adds confusion to the determination of the final position of the wall. The confidence interval of its change will depend on the range of change of the relative error of the computed strength characteristics of the rocks. It will lead to a change in the boundaries of the mine: from α_1 and α_2 in depth H (Figure VI-2). For our example, α_i changes from 30° to 50°, and the depth is $H = 150$ m.

With the selected direction of development of mining works and a slope of working wall of $\varphi_{max} = 15°$, the stage-by-stage volumes of overburden and mineral have been consecutively determined (Figure VI-2). On the basis of these data the increasing volumes of overburden and mineral P_i have been computed out of the depth of development H and the cumulative graphs $V_i = f(P_i)$, $P_i = f(H)$ and $V_i = f(H)$ have been drawn (Figure VI-3).

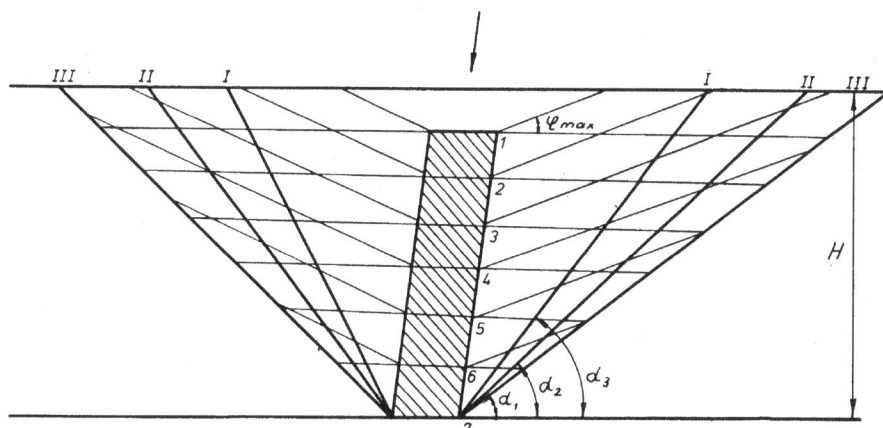

Figure VI-2. Determining the boundaries of open-pit mine exploiting steep deposits by taking into account the probabilistic nature of the length of the sloe of the non-working wall

On the basis of A. I. Arsentiev's method (1987) the values of the average coefficient of overburden n_{av}, of the initial coefficient n_0, of the most adverse operational coefficient of overburden n_i and of the coefficient of non-uniformity of overburden works λ have been determined out of the cumulative schedule:

$$\lambda = \frac{n_i}{n_{av}-n_0},$$

as well as the share of overburden stripped during the construction period of the mine

$$\mu = \frac{n_0}{n_{av}}.$$

The values of these indicators at angles of non-working wall's slope of 30°, 40° and 50° are given in Table VI-3.

The cumulative graphs in Figure VI-3 show that an increase of extraction by P_i and deepening of mining works by h_i corresponds to certain increase in the volume of overburden. Thus, for instance, at $H = 150$ m, $P = 1.0$ million m³, $h = 20$ m and angles of non-working wall of 30°, 40° and 50°, these increases are $V^{30} = 7.8$ million m³, $V^{40} = 4.4$ million m³, and $V^{50} = 2.7$ million m³. In the figure the schedule $V = f(H)$ for $\alpha = 30°$ at an averaged coefficient of overburden is shown in Figure VI-2 with otted line. From these graphs the volumes of overburden which hare additionally stripped or conserved in individual stages can be determined.

Table VI-3. Indicators of a mine at height of 150 and 100 m and at different angles of non-working wall

Wall angle	Wall height, m									
	H=150 m					H=100 m				
	n_{av}	μ_1	n_1	λ_1	n_{pc}	n_{av}	n_1	μ_1	$\lambda\lambda_1$	n_{pc}
	m^3/m^3									
30°	7,55	0,055	13,20	1,85	2,80	5,63	0,125	7,83	1,58	3,33
40°	5,32	0,081	7,20	1,47	3,52	3,90	0,179	4,33	1,53	3,89
50°	3,85	0,112	2,34	1,30	5,17	2,86	0,245	2,67	1,24	4,25

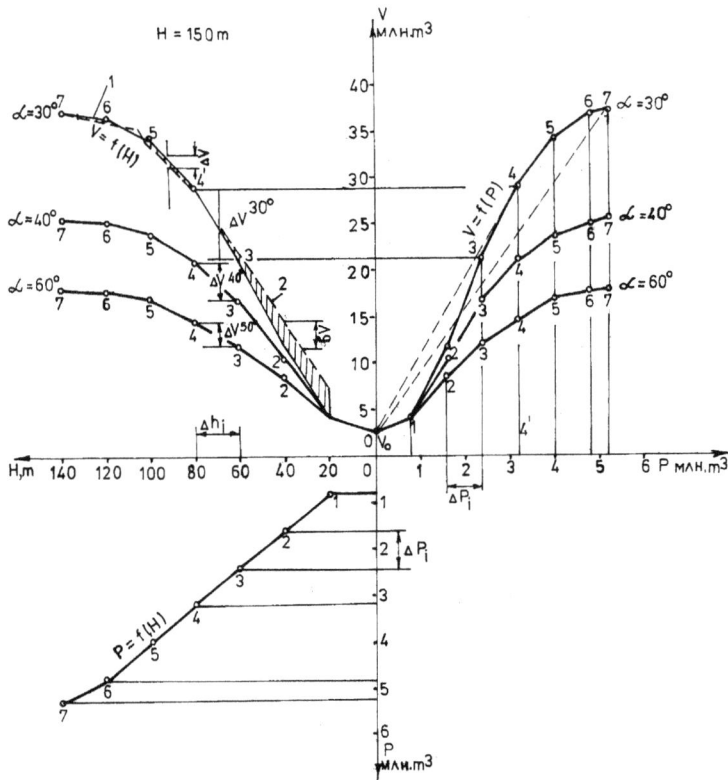

Figure VI-3. Schedule of mine's regime of mining works in case of $H = 150$ m; 1 – before averaging of the overburden coefficient; 2 - after averaging of the overburden coefficient.

The research by S. Hristov (2013) shows that the strength indicators of rocks have a distribution close to the normal one. Thus, we will use the law of normal distribution of errors in the strength indicators of rocks and the error in computing of the average coefficient of overburden will be deemed commensurable with the error in computing the stability of wall's slope.

In case of error in computed strength indicators of rocks $\delta = 0.36$, average squared deviation $\sigma = 0.12$ and probable deviation $E = 0.674$, the degree of risk can be determined in different wall slopes (S. Hristov, 2013).

	$\alpha_i,°$	$R_0,\%$
$tg\alpha_0 = tg\alpha_m(1 - 3\sigma) = 0.576$	30°	0
$tg\alpha_1 = tg\alpha_m(1 - 2\sigma) = 0.680$	34°10'	2.3
$tg\alpha_2 = tg\alpha_m(1 - \sigma) = 0.790$	38°10'	15.9
$tg\alpha_3 = tg\alpha_m(1 - E) = 0.827$	39°50'	25.0
$tg\alpha_m = tg\alpha_0 = 0.900$	42°	50.0
$tg\alpha_4 = tg\alpha_m(1 + E) = 0.970$	44°	75.0

where α_m is the mathematical expectation of wall's angle.

The graphs of relations between the average overburden coefficient and the degree of the risk generated by the increase of the angle of non-working wall's slope are given in Figure VI-4,a. The same figure also shows the change in mine's productivity by mining mass at different wall angles.

We will determine the mine's boundaries by the method of permissible average overburden coefficient (A. Arsentiev, 1987).

$$n_{av} = \frac{n_l}{\lambda - \mu(\lambda - 1)}, \qquad (VI.3)$$

where n_l is the limit coefficient of the overburden (in this case $n_l = 5$ m^3/m^3).

Table VI-3 shows the results obtained from the computation of n_{av}. The point of intersection B in Figure VI-4,a satisfies the equation $n_{av} = n_{pc}$. At that, the angle of wall's slope is $\alpha = 44°$, the mine's productivity $A_{mm}^2 = 7,0$ million m^3, million m^3, and the degree of risk $R = 75\%$, which is practically unacceptable if normal work is to be ensured in the mine. In this relation we will carry out analogous research in case of mine's depth of $H = 100$ m. The results obtained are given in Table VI-3 and in Figure VI-4,b.

If the mine's depth is $H = 100$ m the condition $n_{av} = n_{pc}$ is fulfilled for point B_1 (Figure VI-4,b). In this case the mine is formed at an angle of the non-working wall $\alpha = 40°$, $A_{mm}^2 = 6.5$ million m³ million m³ and degree of risk $R = 25\%$. If the mine needs to operate at other depths as per the schedule in Figure VII-8 the degree of risk can be determined by interpolation. Thus, for instance, if $H = 125$ m and $\alpha = 42°$, the degree of risk is $R = 50\%$.

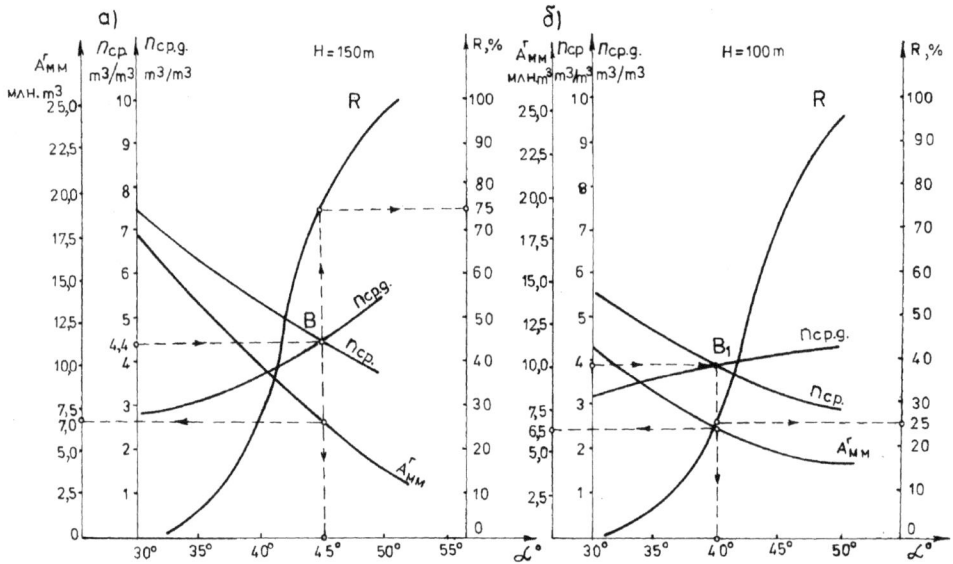

Figure VI-4. Graph of relations between the average overburden coefficient n_{av}, the permissible overburden coefficient n_{pc}, the mine's production capacity A_{mm}^2 and the angle of wall's slope α, by taking into account the degree of risk R; in case of mine depth of: a) $H = 150$ m; b) $H = 100$ m.

The reverse problem can also be solved: to determine the limit depth of the open-pit mine in case that the degree of risk is set. By taking into account the probable negative and positive error in determining the strength indicators for the rocks it is purposeful to determine the mine's boundaries during the design at two values of its depth: the first one with risk lower than 50% i.e. $H < 150$ m and the second one: with risk $R > 50\%$, $H > 150$ m.

When making the final decision on the boundaries of the open-pit mine the economic factors must also be taken into consideration by using the utility function.

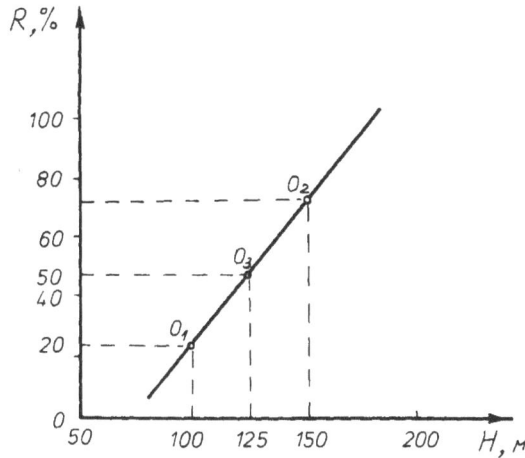

Figure VI-5. Determining the limit depth of an open-pit mine at certain degree of risk

The suggested method makes it possible to use a probabilistic approach when determining the boundaries of the open-pit mine and to take into consideration the degree of risk when assessing the angle of non-working wall's slope. The method enables the reliable determination of the optimal boundaries of the mine. It can be applied in design and development of steep deposits of the type of Elatsite and Asarel.

When taking into account the error while determining the strength indicators of the rocks and the technical and economic data a more proper assessment of mine's boundaries is obtained.

VI.3. STABILITY ASSESSMENT OF SLOPES OF OPEN-PIT MINES

The computation schemes of VNIMI[20] (L. Dimov, 1979) have become established when computing the wall stability in open-pit mines for ore extraction as also confirmed by practice. In these

---

20 Scientific Research Institute of Mining Geomechanics and Mine Surveying of Russian Federation

schemes the sliding surface is made up of a vertical, a sloping and a circular-cylindrical part.

In this case the stability coefficient is computed by the formula (Figure VI-6)

$$\eta = \frac{\sum N \mathrm{tg}\varphi_i + cL}{\sum T} = \frac{\sum_{i=1}^{n} P_i \cos\alpha_i \mathrm{tg}\varphi_i + c_i L}{\sum_{i=1}^{n} P_i \sin\alpha_i}, \qquad (VI.4)$$

where N and T are the normal and the tangential forces on the sliding surface, N/m^2;

φ is the angle of internal friction, °;

c is the cohesion of the rocks, N/m^2;

L is the length of sliding surface, m.

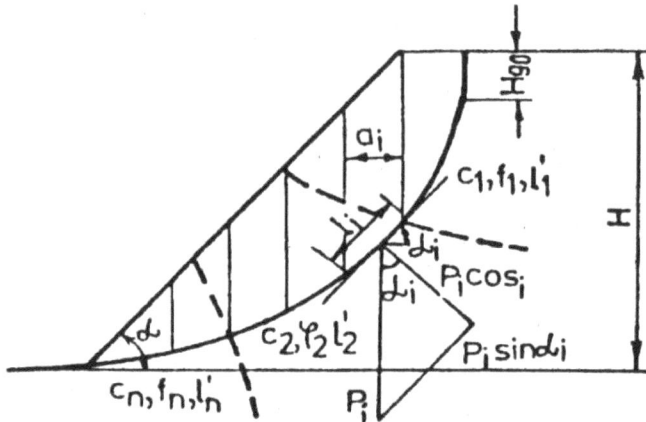

Figure VI-6. Scheme for determining the wall stability

The obtained values of M_η and σ_η determined the probability and the density of distribution of the function (VI.4).

Every decision on slope stability having already been made contains an element of indeterminateness and is associated with certain risk. For that reason the unequivocal characteristic of slope stability as per the traditional criterion, i.e. stability coefficient, is untenable and subjective. A new criterion must be introduced into computations: degree of risk. If used together with the certainty stability coefficient this will enable more accurate assessments of the walls to be made (S. Hristov, 1982).

In this particular case the risk has two varieties. One is associated with the monotonous change of the results. For example, the mine's productivity, the quantity of excavators in the mine, etc. The other risk variety is associated with limit loss. Slope stability belongs to the latter. It might be stable or it might slide down (get caved).

V. Abchuk (1983) suggests the degree of risk to be determined by the difference between one and the probability of the occurrence of certain expected event $P(\varepsilon)$

$$R = 1 - P(\varepsilon), \tag{VI.5}$$

where R is the;

ε is a random quantity, which is beyond a given confidence interval;

$P(\varepsilon)$ is the confidence probability.

To compute the degree of risk associated with the determination of wall's angle the following formula can be used

$$R_1 = \frac{\int_{\alpha,min}^{a_1} f(\alpha)d(\alpha)}{\int_{\alpha min}^{a max} f(\alpha)d(\alpha)}, \alpha_{min} \leq \alpha \leq \alpha_{max}, \tag{VI.6}$$

where $f(\alpha)$ is the density of distribution of wall's angle.

When determining the economic risk the risk coefficient can also be determined by the expression (A. Arsentiev, 1987).

$$K_R = \frac{\int_{\alpha,min}^{a_1} f(\alpha)d(\alpha)}{\int_{\alpha min}^{a max} f(\alpha)d(\alpha)}, 0 \leq K_R \leq \infty, \tag{VI.7}$$

The following relation between the degree of risk and the risk coefficient exists

$$K_R = \frac{R(P_\varepsilon)}{1 - P(\varepsilon)}. \tag{VI.8}$$

In addition to the purely economic assessment of the consequences of risk the psychological (situational) consequences should also be assessed. The results are influenced not only by objective but also but by subjective factors because the computations are made not only by machines but also by people.

J. Kemen, J. Thompson (1961) suggested eight options for assessment of the advantages and disadvantages of error in decision making. When examining the problem of slope stability management it is useful to adopt the function of concerns for consideration $\eta(\delta)$.

In his research A. Arsentiev (1987) suggested that only 4 options for assessment of decisions should be used in mining.

1. Brave attitude to risk: the function of concerns is $C(\delta) = a(1 - e^\delta)$, a is a coefficient of proportionality (price per unit of relatively increased (decreased) amount of the parameter δ. being examined.

2. Calm (balanced) attitude to risk: the function of concerns is $C(\delta)_b = a\delta$.

3. Careful (cautious) attitude to risk: $C(\delta) = a(e^\delta - 1)$.

4. No risk $C(\delta) = 0$.

The selection of possible risk depends on many diverse factors: scarcity of mineral resource, or electricity and metal, respectively; speed of development of mining works; financial condition of the sector; type of machinery and facilities and the possibility of buying such, etc. For example during the period of operation of the mine it becomes necessary to increase the production capacity or to decrease the number of excavators by overburden or to decrease the volume of the overburden because of deteriorated technical-economic indicators. This can be achieved in the most rapid way only if the angle of the wall is increased. At the same time, however, we are not sure that there will be no landslide and that the joint-stock company or the sector will sustain no loss.

In such case one of the typical tasks for the application of the utility function (V. Abchuk, 1983) is to obtain profit in case of a possible landslide of the block. We will examine the essence of this task by means of a simple example. Let us assume that a landslide occurs if the angle of the wall is increased and the sum €A will have to be spent or order to resume the mining works. If, however, no landslide occurs a large sum €B will be gained due to the non-excavation of overburden, transportation, etc. and this will also cover the funds that would have been invested for eliminating the consequences of a possible landslide of the wall.

Let us designate the probability of profit in the absence of landslide of the wall by the letter P and find what the amount of the profit for the undertaking should be. It consists of two parts:

- If a landslide occurs, a sum equal to € $p(-D)$ will be lost.

- If no landslide of the wall occurs the sum of € $(1 - p)A$ will be gained.

Therefore, one runs the risk only in the case where the total sum is positive

$$B_{com} = P(-D) + (1 - p)A > 0;$$
$$-pD + A - pA > 0; A > p(D + A).$$
(VI.9)

For instance, if the probability of profit p is 10% and the inequality $0.1 < \frac{A}{A+B} AA + B$, is observed, this means that if the value of profit is €5 million the value of elimination of the landslide will be €559 thousand.

However, we are asking how to proceed if it becomes necessary to increase the production capacity while preserving the existing machines. If the angle of the wall is increased, doubts arise over the occurrence of a landslide. Is it worth it to run the risk in this case? The reasoning is as follows: if there is no landslide, profit will be equal to pD, and if a landslide occurs the loss is $(1 - p)(-A)$.

Therefore, the total sum of profit will be

$$B_{com} = pD + (1 - p)(-A) = pDpA - A$$
$$= -A + p(D + A)$$
(VI.10)

i.e. it is profitable to run the risk if $-A + p(D + A) > 0$ or $A < p(D + A)$.

However is it always worth it to run the risk? It depends on the particular case, i.e. on the ratio of profit if there is no landslide to the loss. The concept of risk utility is used to clarify this case (V. Abchuk, 1983).

Let us construct a graph of the choice: to run or not to run the risk in case of a possible occurrence of a landslide (Figure VI-7). The value of the expenses to be spent to eliminate the landslide are entered on the abscissa from 0 to 1 in conditional units, and the probability of profit from saved volumes of overburden are entered on the ordinate, also in conditional units from 0 to 1. The monetary profit in this case is assumed to be equal to one conditional unit.

In case of equal attitude to risk the utility of profit is

$$C(B_t) = B_t = -A + p(D + A). \qquad \text{(VI.11)}$$

The utility function for equal attitude is

$$p = \frac{A}{A + B}.$$

In case of careful attitude the utility of the overall result is

$$\begin{aligned} C(B_t) &= C[pD + (1 - p)(-A)] \\ &= C(pD) + C((1 - p)(-A)) \qquad \text{(VI.12)} \\ &= p(1 - e^{-D}) = (1 - p)(1 - e^{-A}). \end{aligned}$$

The risk makes sense if

$$C(B_t) = p(1 - e^{-D}) + (1 - p)(1 - e^{-A}) > 0, \qquad \text{(VI.13)}$$

or

$$p > \frac{e^A - 1}{1 - e^{-D}}.$$

On the graph this line corresponds to the dotted line and two points (Figure VI-7) and shows the marginal position of

$$p = \frac{e^A - 1}{1 - e^{-D}}.$$

In case of brave attitude the utility is

$$C(B_t) = p(e^D - 1) + (1 - p)(e^{-A} - 1) > 0$$

or

$$p > \frac{1 - e^{-A}}{e^D - e^{-A}}.$$

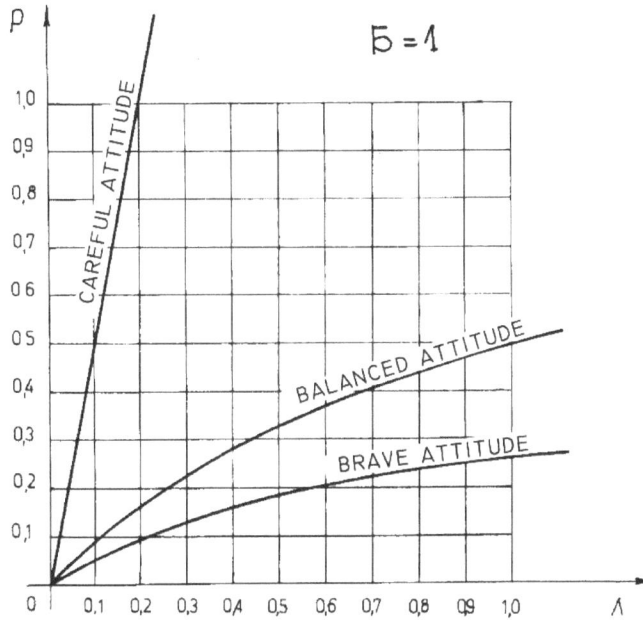

Figure VI-7. Graphs of utility in case of careful, balanced and brave attitude to risk

In Figure VI-7 this line is indicated by a point and a dotted line and shows the marginal position of

$$p = \frac{1 - e^{-A}}{e^D - e^{-A}}$$

and denotes the range of permissible risk in case of brave attitude.

The graphics shows that it is not always purposeful to run risks. For example if the probability of profit is 0.1, risk should only be taken if the value A (the sum spent to eliminate the landslide) is in the range from 0.02 to 0.22 conditional units.

This means that if Б is equal to €550,000 the expenses incurred will only be determined if their value is in the range from €11,000 to €121,000.

As evident from the above reasoning the decision as to whether the risk should be taken will depend much on the nature of the utility function.

The decision to take the risk must be made only if the total result is positive.

From the analysis of risk situations and the graph in Figure VI-7 it can be concluded that it is purposeful that the degree of risk should also be taken into account in decision making.

VI.4. THE MONTE CARLO METHOD & THE FINITE ELEMENT METHOD

The main factors that have impact on the stability of mine workings are geological, hydrogeological, physico-mechanical, climatic and mining-technical. We will pay special attention to the influence of physico-mechanical properties on the tense and deformed condition of the massif by the finite element method.

In mathematical models by which this condition is being described these factors participate with their numerical values.

When carrying out engineering work in the rock massif the behavior of rocks must be known, especially in the vicinity of mine workings. The great variety of properties in the rock massif due to the impact of a number of natural phenomena makes it necessary to study them. The characteristics of the massif are studied either under the conditions of its natural occurrence, or based on appropriately selected samples.

In contemporary engineering science and practice the methods using the apparatus of the theory of elasticity find wide application (M. Mazhdrakov, D. Benov, G. Trapov, 2013); in that theory the solving of special problem is reduced to determination of the components of shifts u, v, w, of stress $\sigma_x, \sigma_y, \sigma_z, \sigma_{xy}, \sigma_{xz}, \sigma_{yz}$ and of strain $\varepsilon_x, \varepsilon_y, \varepsilon_z, \gamma_{xy}, \gamma_{xz}, \gamma_{yz}$ in any point of the object being examined. To that end a system of 15 differential equations must be prepared in order to be solved together to enable the determination of real object's stressed and strained state.

In most cases, however, the geometric configuration of objects and the marginal conditions are so complex that the determination of shifts, stresses and strains by solving the respective differential equations turns out to be quite difficult, and sometimes even impossible task. The pursuit of higher adequacy of mathematical model and real object leads to taking into account a number of factors, which could hardly be dome within the scope of the classical theory of elasticity. This requirement has lead to the development of numerical methods and, in particular, to the widespread use of calculus of variations in rock mechanics.

The main task of calculus of variations is to find a function for which the functional of the total potential energy of the system expressed as a function of the shift reaches an extremum.

Under the finite element method it comes to solving a system of linear equations, which can be written in matrix form thus

$$[K]\{u\} = \{F\}. \tag{VI.14}$$

This equation is the main equation of the finite element method and its solution is

$$\{u\} = [K]^{-1}\{F\}. \tag{VI.15}$$

The matrix $[K]_{3m,3m}$, where m is the number of nodes in the network is a generalized stiffness matrix of the system. Its elements are determined by the geometry of nodes (their coordinates) and by the deformation characteristics of the environment under examination. There are different ways of finding the generalized matrix of system's stiffness. Relatively the easiest way is to find the stiffness matrices $[K]_e$ of individual elements and then to address their members and add them to their respective elements of the generalized stiffness matrix $[K]_{3m,3m}$ of the system.

The elements of matrix's column $\{u\}$ are shifts, and those of $\{F\}$ are the active forces in every node of discretization network.

To illustrate the ideas about the application of the finite element method under conditions of changeability of the rock properties, here a homogeneous and isotropic environment is examined as in case of discretization of the environment a triangular element is used.

A model that takes into account the random nature of rock properties and of quantities being functions thereof. The values of rock properties change in pace and certain values of these indicators in the particular case are realized with some probability. All of this shows the clearly manifested probabilistic nature of rock properties.

As the physico-mechanical properties are random quantities all non-random functions with these arguments are also random quantities.

For the sake of greater clarity of the text it is assumed that ρ is a constant and only E and μ are random quantities. The combination of the two components (E, μ) is a two-dimensional random quantity. As a result of the observations made (laboratory tests) a sample has been drawn with volume n for the dimensional random quantity (E, μ) and its empirical law of distribution has been established. In fact, this is the joint distribution of (E, μ). Below, E values are designated by x, and μ values − by y. The joint function of distribution of (E, μ) will be denoted by $F(x, y)(E, \mu)$.

According to the formula

$$F(x, y) = F_x(x) F_{yx}\left(\frac{y}{x}\right), \qquad (\text{VI}.16)$$

the imitation of the quantity (E, μ), can be consecutively obtained by means of a random pair η_1, η_2 independent, uniformly distributed in the interval $[0; 1)$ random numbers. To that end the inverse functions of $F_x(x)$ and $F_{yx}\left(\frac{y}{x}\right)$ have been used in the following sequence. The value $x^* = F_x^{-1}(\eta_1)$. is obtained by means of the established number η_1 and the inverse function F_x^{-1} x^* shows which is the corresponding conditional function of distribution $F_{yx}\left(\frac{y}{x^*}\right)$. By using it the realization y^* of the process at the second step under the formula $y^* = F_{yx}^{-1}\left(\frac{\eta_2}{x^*}\right)$. is determined by the established number η_2 and the inverse function F_{yx}^{-1}. Thus, one imitation of the quantity (E, μ) has been made.

This process is performed repeatedly in order that sufficient number of such realization is obtained as in each of them the stresses and strains in certain nodes are obtained by means of the established shifts. The aim is to use them to find assessments of the characteristics sought.

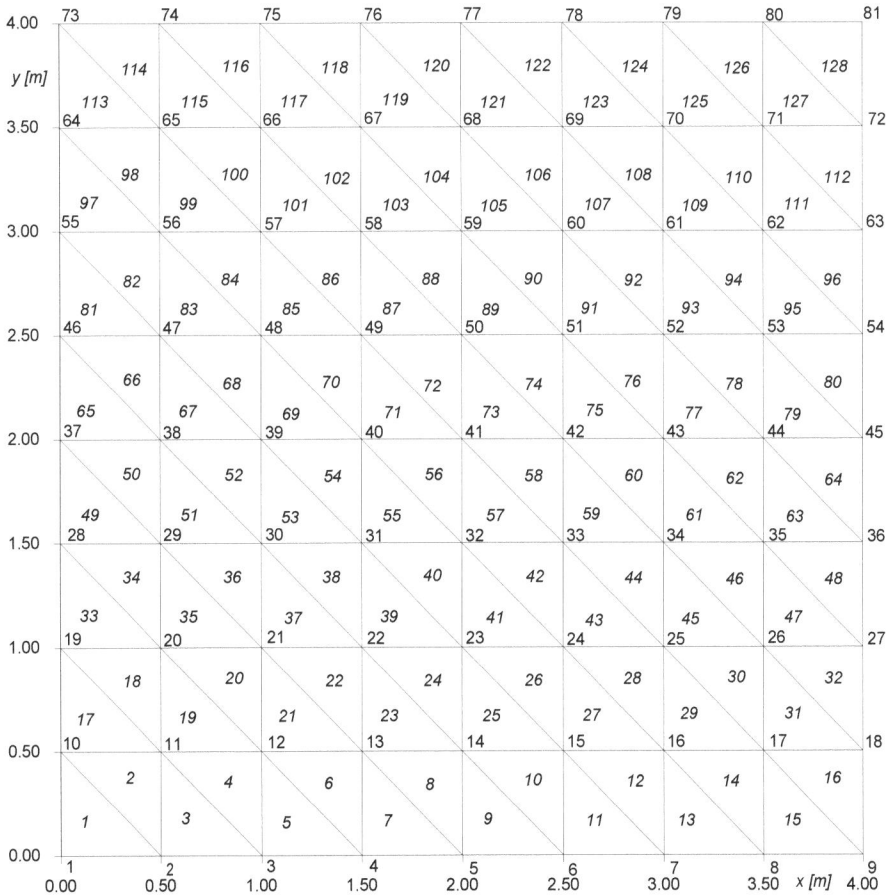

Figure VI-8. Scheme of discretization for the "deep beam" example"

Machine realization of the model. Input data is the number of realizations of imitation and the empirical distributions for E, μ and ρ from laboratory experiments.

Computations are made in the following sequence.

1. The input data are read.

2. The table of joint interval statistical distribution and the corresponding cumulative function are established by using the sample of the three random quantities E, μ and ρ.

3. Three independent, uniformly distributed in the interval $[0; 1)$ random numbers η_1, η_2 and η_3, are obtained by using a random number generator, and by these numbers the random quantity (E, μ, ρ) is imitated in the aforesaid manner.

4. The serial realization of the quantities $\{\varepsilon\} = \left\{\varepsilon_x, \varepsilon_y, \varepsilon_{xy}\right\}^T$ and $\{\sigma\} = \{\sigma_x, \sigma_y, \sigma_{xy}\}^T$ is obtained by using the computing module for application of the finite element method $\{\varepsilon\} = \{\varepsilon_x, \varepsilon_y, \varepsilon_{xy}\}^T$ and $\{\sigma\} = \{\sigma_x, \sigma_y, \sigma_{xy}\}^T$.

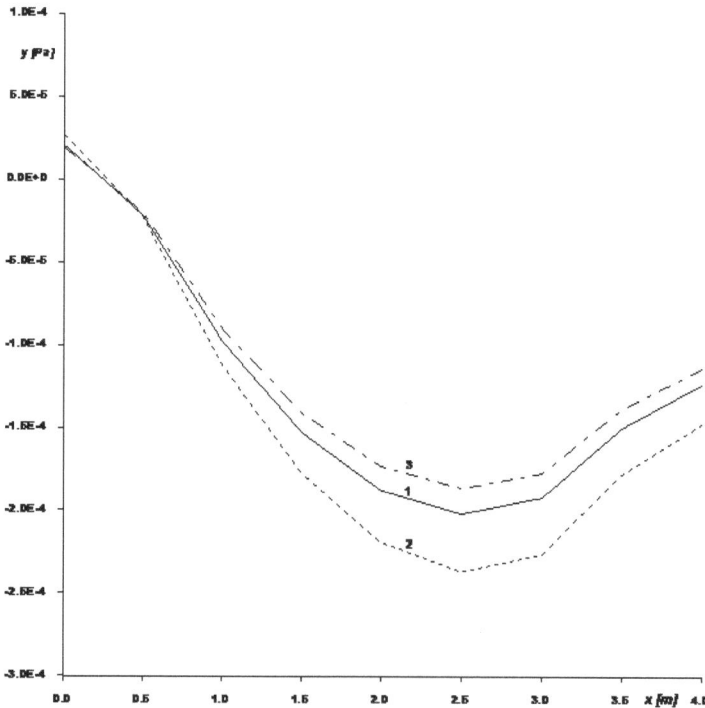

Figure VI-9. Stresses σ_y along the "19-27" section

Items 3 and 4 are performed the set number of times and then it is proceeded to item 5.

5. The desired assessment of $\{\varepsilon\}$ and $\{\sigma\}$ is obtained and the results are exported to an appropriate external device.

Results and conclusions. As an example we will cite the study of the behavior of the "deep wall" (Figure VI-8), which is examined as a massive, homogeneous and isotopic object. The discretization of environment was made by 128 triangular elements. After repeatedly running the computation procedure the shifts, stresses and strains were obtained. The values obtained for the stresses σ_y taken along the "19-27" section connecting the nodes with numbers 19 and 27 are used as an illustration (Figure VI-9).

The application of the finite elements method under conditions of changeability of the rock properties enables the results of the natural and laboratory observations to be reflected more fully, which increases the adequacy of the mathematical model. The realizations made of the model provide grounds to maintain that it can be successfully considered to build on the traditional application of the finite element method.

REFERENCES

[1] Abchuk, V. (1983). <u>Risk theory in the marine practice</u>. Leningrad, Sudostroenie [in Russian].

[2] Arsentiev, A. (1987). <u>Modern principles of the theory of quarrying</u>. Leningrad, Nauka [in Russian].

[3] Dimov, L. (1979). <u>Mine Surveying Handbook</u>. Sofia, Tehnika [in Bulgarian].

[4] Hristov, S. (1982). One method for analyzing and estimating the coefficient of stability of slopes in opencast mines. <u>Vaglishta</u>. Sofia. 2 [in Bulgarian].

[5] Hristov, S. (2001). <u>Drainage and resistance to opencast mines and quarries</u>. Sofia, MGU St. Ivan Rilski [in Bulgarian].

[6] Hristov, S. (2013). <u>Technological and geomechanical problems in the design and exploitation of opencast mines and quarries</u>. Sofia, MGU St. Ivan Rilski [in Bulgarian].

[7] Kemen, J., J. Thompson (1961). The influence of psychological attitudes on game outputs. <u>Matrix games</u>, Fizmatgiz [in Russian].

[8] Mazhdrakov, M., D. Benov, N. Mihaylov (2012). <u>Risc management possibilities of sliding processes in open pit mines</u>. Third National Scientific and Technical Conference with International Participation "Technologies and Practices in Underground Mining and Mine Construction", Devin, Bulgaria.

[9] Mazhdrakov, M., D. Benov, G. Trapov (2013). <u>CADMin. Automated Planning of Development of Mining Works in Open Pits</u>. Sofia, MGU St. Ivan Rilski [in Bulgarian].

[10] Zlatanov, P., M. Mazhdrakov, G. Trapov (1985). "Influence of the possible deviations of the physico-mechanical values of the lithological varieties on the stability of the working panels in the East Maritza site." <u>Minno delo</u> 12: 4-6 [in Bulgarian].

SECTION VII.
PRODUCTIVITY OF MACHINERY IN OPEN-PITS

M. Mazdrakov, St. Hristov, D. Benov, Iv. Ivanov, V. Shishkov

VII.1. RISK IN BUCKET-WHEEL EXCAVATORS' OPERATION IN LANDSLIDE DEFORMATIONS

Landslide deformations and landslides are a big problem in coal extraction in open-pit mines where heavy-duty excavators work on the overlying complex and coal layers under complex geotechnical conditions. Different defects (cracks), which are activated upon the operation of excavators have a crucial role. By their genesis, the cracks are:

— lithification, micro- and meso-cracks situated in a netlike form in the massif (Q_I);

— meso-and macro-cracks of diagenetic and tectonic origin (Q_{II});

— mirror macro-crack surfaces (Q_{III}).

Upon the occurrence of landslides the process is carried out as per the scheme of active earth pressure prism, central block and passive prism (Figure VII-1, a). When the lithification cracks (defects) and mobilized the active earth pressure prism is shaped on the formation surface (Figure VII-1, b).

The surface is wavelike due to the netlike distribution of micro- and meso-cracks in the total volume of overlying and coal complex. Macro-cracks, especially in coal, lead to a block break-off.

Mirror surfaces in the base of the walls are related to their deformation. These surfaces are inclined towards the worked-out space or toward the massif or are horizontal.

Table VII-1. Average values and confidence intervals for embankment's stability coefficient.

Scheme	Number of tests	Average value of F	95% confidence for F	
			left end	right end
A	20K	0,92	0,77	1,07
A	50K	0,91	0,77	1,06
A	100K	0,91	0,78	1,04
B	20K	1,14	1,00	1,28
B	50K	1,13	1,02	1,25
B	100K	1,13	1,03	1,23

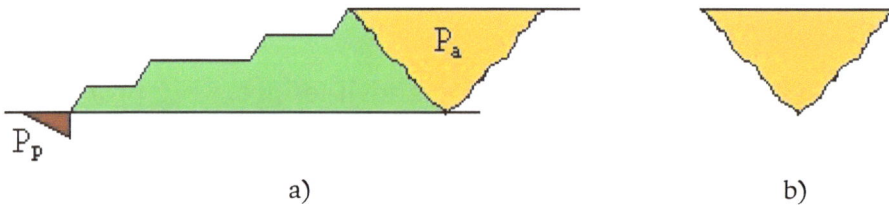

a)   b)

Figure VII-1. Location of the active pressure prism

The defects Q_I, Q_{II} and Q_{III} being examined are characterized by their empirical laws of distribution.

The solution to the problem of assessing the technological risk in the operation of large bucket-wheel excavators is complex in nature. Bucket-wheel excavators operate in the intact massive of the overlying complex and coal layers. The excavators often get close to the active earth pressure prism, which is inclined to rheological deformations. The vibrations of excavators on the worksites mobilize the defects in different ways.

For the assessment of the technological risk in bucket-wheel excavators a probabilistic model, by which values of Q_I, Q_{II} and Q_{III}, and of properties c and γ (it is assumed that the angle φ is practically constant) are obtained by means of stochastic modeling, is applied.

This assessment makes it possible to make an actual judgment of the risk and of the increase in possible revenues in case of an adaptive management of the stability of mine's wall.

VII.2. OPTIMAL NUMBER OF DUMP TRUCKS LOADED BY ONE EXCAVATOR

Transportation of mined mass is one of the main technological processes in open-pit mines. The general estimation is that transportation expenses make up between 20% and 40% of the cost of extracted mineral. In particular, in case of automobile transport the expenses are definitely high: about 35 to 37%, depending on dump trucks' type and capabilities (A. Atanasov, P. Zlatanov, D. Stoyanov et al., 2001).

The number of dump trucks that work with one excavator depends on the time for performance of one run. This time is determined by two types of indicators: ones that do not depend on the number of dump trucks, and ones that do (M. Mazhdrakov, D. Benov, V. Shishkov et al., 2014).

The independent indicators are: the loading time t_l and the unloading time t_{unl}, time for maneuvers upon loading t_{lm}, and upon unloading t_{unlm}, which do not influence the remaining dump trucks.

The operating times of loaded/empty dump truck t_{loaded} and t_{emp} are relatively independent.

The waiting (queuing) times upon loading t_{lq} /upon unloading t_{unlq} are dependent on the number of dump trucks.

The total time of one run (cycle) is

$$T = t_{lq} + t_l + t_{lm} + t_{loaded} + t_{unlq} + t_{unl} + t_{unlm} + t_{emp}, \min. \tag{VII.1}$$

If we assume that the first dump truck is partially loaded the waiting time in case of a queue of n_0 dump trucks is

$$t_{lq} = (n_0 - k)t_l, \min, \tag{VII.2}$$

where k is a coefficient that takes into account the extent of loading $(0 \le k \le 1)$.

The same is also valid for unloading:

$$t_{unlq} = (n_0 - k)t_{unl}, \text{min.} \qquad \text{(VII.3)}$$

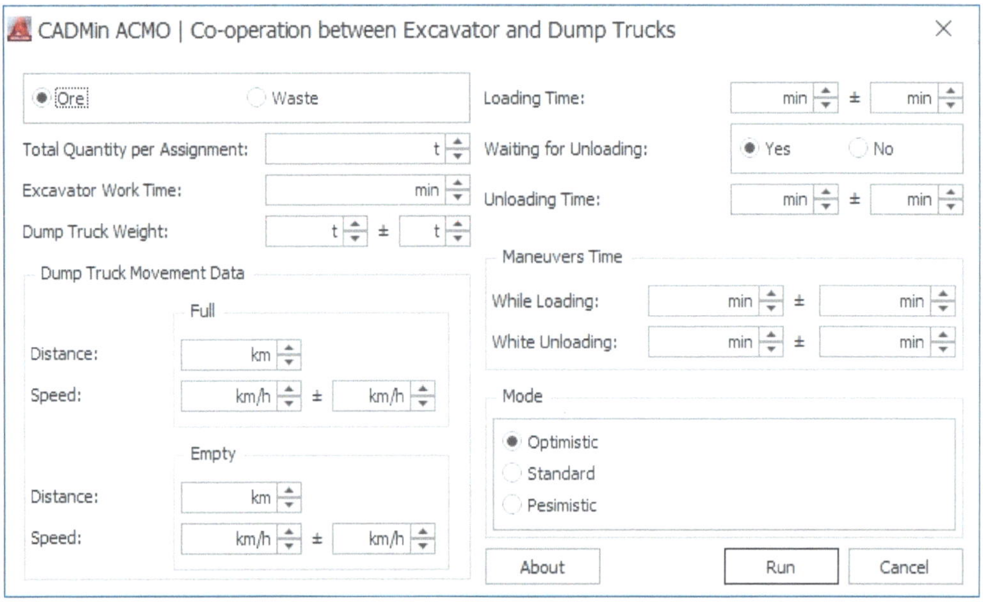

Figure VII-2. Joint operation of an excavator and dump trucks, CADMin (I. Ivanov, G. Royalski, D. Benov, 2017)

The queue length is a probabilistic quantity that depends on the number of dump trucks. The probability of a dump truck being lined up in the loading/unloading queue is

$$p_{l,1} = \frac{t_{lq} + t_l}{T},$$

$$p_{unl,1} = \frac{t_{unlq} + t_{unl}}{T}. \qquad \text{(VII.4)}$$

If 4 dump trucks with numbers 1, 2, 3 and 4, respectively, are in operation, in front of No. 4 there may be a queue:

- of 3 dump trucks, 1 combination: numbers 1, 2 and 3, with probability p^3 and $(3 - 0.3)t_{l/unl}$ waiting time;

- of 2 dump trucks, 3 combinations: numbers 1 and 2; 1 and 3; 2 and 3, with probability waiting time p^2 and waiting time $(2 - 0.3)t_{l/unl}$;

- of 1 dump truck, 3 cases: numbers 1, 2 or 3, with probability p^1 and $(1 - 0.3)t_{l/unl}$ waiting time.

The case where there is no queue in front of the dump truck is also possible, most often in the case of a long transportation distance and a small number of dump trucks; then the waiting time is 0.

With the times chosen by the formula (VII.1) firstly we get an approximation of T and recurrent values of t_{lq} and t_{unlq} (VII.2). After certain number of iterations a sufficiently accurate value of the expected duration of one run is obtained T. Then, each dump truck will perform

$$n_k \leq \frac{T_\Sigma}{T}, runs, \qquad (VII.5)$$

and will carry $n_k q$ units of mined mass. To perform the task of Q units of mined mass there will be needed

$$n_a \geq \frac{Q}{n_k q}, dump\ trucks. \qquad (VII.6)$$

The nor equals sign in the formulae (VII.5) and (VII.6) reflects the specifics of the integer solution. The productivity of dump trucks will be re-dimensioned by

$$d = n_a n_k q - Q. \qquad (VII.7)$$

The efficiency of the solution is inversely proportional to the value of the remainder d.

1. The total time spent by dump trucks in queue and in "idle" operation

$$T_0 = n_c n_k (t_{то} + t_{по}). \qquad (VII.8)$$

Regardless of the accepted algorithm, the problem under consideration has a solution only when the total time is sufficient to load the planned mining mass, which is expressed by the inequality.

$$\frac{Q}{q} t_l < T_\Sigma, min. \qquad (VII.9)$$

We solve the aforesaid problem by the Monte Carlo method. The behavior of the model is studied by changing the indicators $Q, q,$

t_l, t_{lm}, t_{unl}, t_{unlm}, t_{loaded}, t_{emp}. The efficiency of the solution is presented by the distribution of the values of the difference d (VII.7) (Figure VII-3) and the total duration of on-site operation (VII.8) (Figure VII-4).

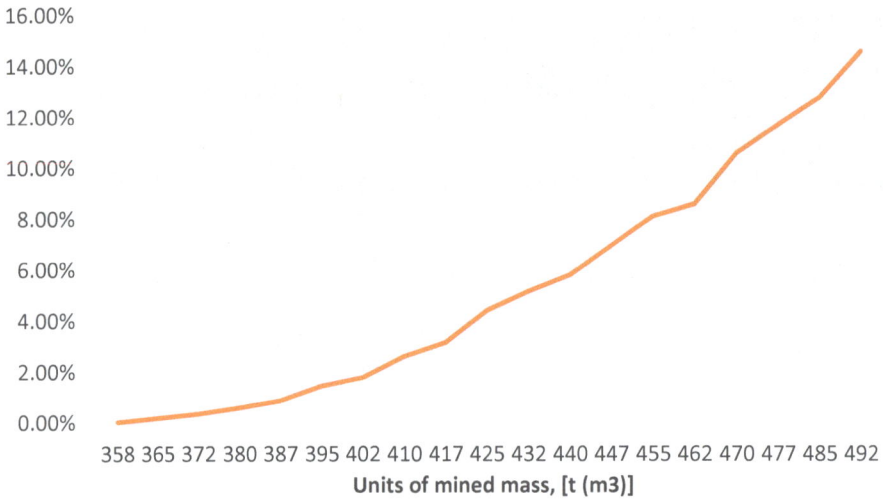

Figure VII-3. Distribution of risk upon re-dimensioning

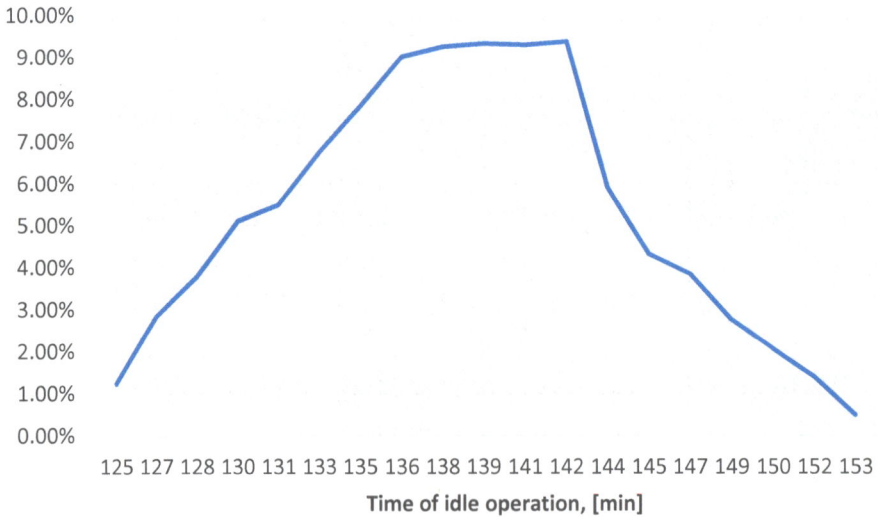

Figure VII-4. Distribution of aggregate time of idle operation

The solution obtained is an integer, whereupon the specialist can choose between:

– partial utilization of transport opportunities (rounding up to the nearest integer);

– operation at certain risk (rounding down to the nearest integer).

The indicator of aggregate time of idle operation of dump trucks (VII.8) (Figure VII-4) is used for the assessment of the solution.

The input parameters are determined by processing the shift reports (S. Stankov, 2001) or by special studies (M. Mazhdrakov, B. Benova, 2002).

REFERENCES

[1] Atanasov, A., P. Zlatanov, D. Stoyanov, et al. (2001). Open-mining technology. Sofia, MGU St. Ivan Rilski [in Bulgarian].

[2] Ivanov, I., G. Royalski, D. Benov (2017). ACMO Module - information system for structuring, maintaining, organizing and storing the engineering information in the "Ellatzite" open pit. XIV International Conference of the Open and Underwater Mining of Minerals, Varna [in Bulgarian].

[3] Mazhdrakov, M., D. Benov, V. Shishkov, et al. (2014). Determining the Optimum Number of Dump Trucks Loaded by One Excavator in Open Pit. Fourth National Scientific and Technical Conference with International Participation "Technologies and Practices in Underground Mining and Mine Construction". Devin, Bulgaria: 84-87 [in Bulgarian].

[4] Mazhdrakov, M., B. Benova (2002). Useful capacity and actual cargo volume of dump trucks. IXth Mine Surveying Conference, Varna [in Bulgarian].

[5] Stankov, S. (2001). Optimal Management of Open Pit Transport with GPS NavStar. VIth International Conference of the Open and Underwater Mining of Minerals, Nessebar [in Bulgarian].

SECTION VIII.
MINE SURVEYING AND GEODESY

M. Mazdrakov, D. Benov, Iv. Ivanov,
Il. Ivanova, N. Valkanov

Currently, the preliminary assessment (forecasting) of the expected error in mine surveying and geodesic measurements is a matter of academic interest, and not a wide-spread practice. In our opinion this is because of reasons such as:

— the application of high-precision and reliable tools and well-mastered measurement technologies;

— the acquired professional experience and confidence of specialists;

— the greater accuracy of geodesic base.

Wide application of electronic means of angle and length measurement and computer technologies has increased the specialists' confidence and has increasingly "removed" the accuracy forecasting from day-to-day practice.

There are, however, engineering problems where the pre-assessment of accuracy is a matter of some interest. Such problems are cutting long mine workings along a run with contrary faces or ones to get to a set point, measuring the vectors of removal of control reference points of facilities in charge, settling disputes on stripped volume of mined mass, etc.

The overview of known forecasting models shows some weaknesses related to their methodological base: the law of propagation of error.

VIII.1. ERROR PROPAGATION LAW

Geodesy is one of the few sciences that officially recognize that errors can be made. Moreover, such errors are the object of numerous and thorough studies, which are often designated as "Theory of Errors".

Errors in geodesic measurements can be classified by different signs depending on:

- source: instrumental, personal (observer's) and ones due to the impact of the environment;

- time of measurement: a priori (before measurement) and a posteriori (after measurement).

In error theory errors are divided not by sources but by properties and regularities of occurrence. In terms of the latter signs errors are divided into grave, systemic and random ones.

Grave errors have values that far exceed the expected accuracy of measurements. **Systemic** errors are ones which preserve their sign and value in the measurement process or change subject to certain law.

Random errors vary in size and sign even in case of a relative constancy of conditions upon measurement. These errors are due to imperfections of observer's senses, inaccuracies in workmanship and settings of measuring devices, etc. The experience gained in numerous measurements confirms that random errors have the following main properties:

- under given conditions the absolute value of random errors does not exceed certain limit;

- errors of low absolute value appear more often than the ones of greater value;

- positive and negative errors of same absolute values appear equally often, i.e. they are equally probable;

- by increasing the number of measurements and under a relative constancy of conditions the arithmetic mean of random errors in the measurement of the same quantity tends to zero (more precisely, a zero of negligibly low value is distinguished).

In indirect (mediate) measurement the result is computed from the data obtained upon direct measurement of quantities by which the quantity to be found is connected with mathematical relations. Errors made in direct measurement of input quantities influence the accuracy of final results as the influence of such errors is determined by the type of relation between the quantities measured and the ones computed and the place where the error is made.

In geodesic measurements the impact of errors in output measurements on the accuracy of the result is determined by the law of propagation of error. According to this law the mean squared error of the function of measured independent quantities

$$y = f(x_1, x_2, x_n)$$

is

$$m = \pm \sqrt{\left(\frac{\partial f}{\partial x_1}\right)^2 m_1^2 + \left(\frac{\partial f}{\partial x_2}\right)^2 m_2^2 + \cdots + \left(\frac{\partial f}{\partial x_n}\right)^2 m_n^2}, \quad \text{(VIII.1)}$$

where $\frac{\partial f}{\partial x_1}, \frac{\partial f}{\partial x_2}, \frac{\partial f}{\partial x_n}$ are the partial derivatives of the function with respect to $x_1, x_2, \ldots, x_n$;

$m_1, m_2, \ldots, m_n$ – the mean squared errors, by which $x_1, x_2, \ldots, x_n$ have been measured.

The $\pm$ sign shows that the positive and negative errors are equally probable.

The law of propagation of error is used most often for preliminary (a priori, estimated) computation of error in geodesic works. When interpreting the obtained values of errors we must also take into consideration the following.

1. Although the law has been deduced as a consequence of probabilistic nature of errors in measurements a determined assessment, which meets some definite values of errors $m_1, m_2, ..., m_n$, is obtained by using the law (VIII.1)$m_1, m_2, ..., m_n$.

2. A normal distribution of errors in measurements is presumed.

3. As a rule, various assumptions are accepted in order to obtain a formula suitable for computation, e.g. errors in input quantities are assumed to be identical, some components of the formula (VIII.1) are ignored, etc.

4. In many cases the relation between the input quantities and the final results is very complex and cannot be practically expressed by a formula (VIII.1).

For these reasons we think that the use of the Monte Carlo method is also suitable for the pre-assessment of the accuracy of geodesic and/or mine surveying measurements (M. Mazhdrakov, T. Trendafilov, 1981; M. Mazhdrakov, D. Benov, I. Ivanov, 2012b).

VIII.2. ASSESSMENT OF ACCURACY IN OPEN TRAVERSE

Open traverse starts from a point and a side of known coordinates or a specified angle, respectively, and it advances by measurement of lengths and angles without getting included in a certain terminal point.

The open traverse does not meet some fundamental principles of performance of geodesic works. It does not follow the scheme of development of geodesic meshes: from the general to the special, quite on the contrary: it evolves from the special to the general. In addition, only the minimum number of measurements is performed, i.e. there are no "redundant" measurements, which does permit objective control and assessment of achieved accuracy. For these reasons geodesic statutory documents does not permit this type of traverse upon the development of work base.

In certain cases, however, the application of a open traverse is the only possible solution. Such are mine surveying works in underground cutting of mine workings and transport, hydrotechnical and other tunnels. In such cases, the open traverse follows the digging of the said facilities and at the same time it serves both for tracing of their axis and for surveying of the actual condition of the face (M. Mazhdrakov, D. Benov, I. Ivanov, 2012a).

This task is especially responsible when digging underground galleries and tunnels with contrary faces or, where they must pass through points of known coordinates.

Thus, in Technical Mine Surveying Instruction (L. Cheshankov, P. Ganchev, S. Harizanov et al., 1969), it is recommended in case of workings of great length and/or complex shape to perform a preliminary assessment of accuracy, which is expected to be achieved by means of the tools and measurement methods used.

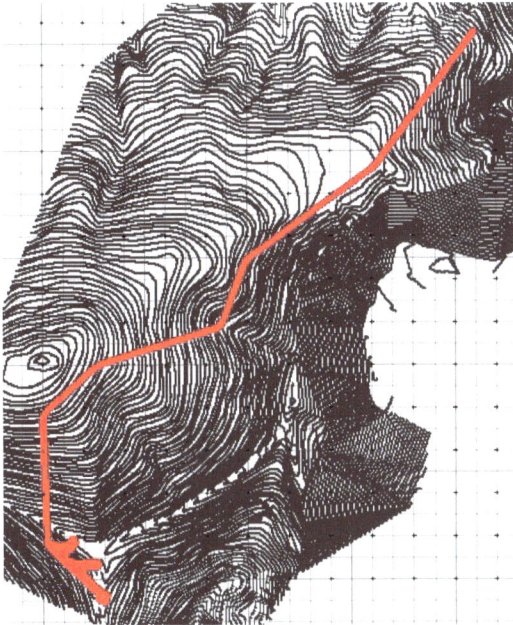

Figure VIII-1. Open traverse that must get to a point of defined coordinates; the length of the traverse is 1868 m

In mine surveying literature it is popular to compute the error in the last point of a open traverse when the polygon's geometric form and the mean squared errors of the lengths and angles are set , based on the law of propagation of error (I. Bahurin, 1936; K. Velev, 1965).

In the practical application of this method two circumstances must be taken into consideration.

1. The volume of computations is considerable. This circumstance was important in the period when the method was created (I. Bahurin, 1936). For this reason approximate formulae have been developed on the basis of certain assumptions: strained traverse, same lengths of the sides, etc. Of course, currently it is not difficult to make the respective computing software program.

2. In our opinion, another circumstance is more important. In the literature it is not pointed out that despite the use of probabilistic formula apparatus the method is in fact determined. According to this method, in case of the same input data, the same deviation of end point is always forecasted. This means that in case of repeated measurement of a traverse by the same tools and by using the same methodology one must always expect to obtain the same coordinates of the end point, which obviously is not true.

For the aforesaid reasons we suggest a solution to the problem by the Monte Carlo method.

The input data include the geometric scheme of the traverse (Figure VIII-1) and the presumable errors of starting parameters: the coordinates of a point and a specified angle and of the measured quantities: polygon angles and lengths.

Two alternatives for the "random number – measurement error" transition are envisaged:

– a normal law of distribution, which is underlying for the geodesic (mine surveying) assessment of accuracy; this alternative is designated "optimistic";

cystemic nature: only "positive" or only "negative" values of errors, thus producing the so-called "limit error"; this alternative is designated "pessimistic".

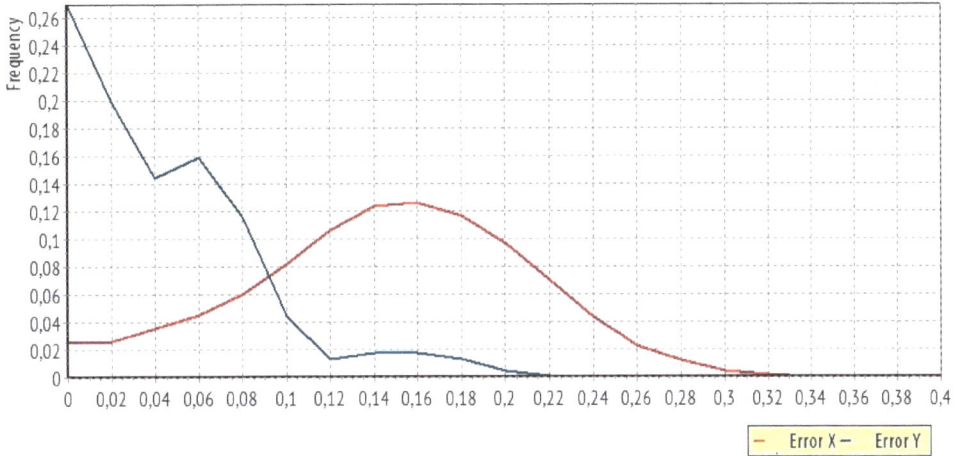

Figure VIII-2. Expected distribution of mean squared errors of coordinates X and Y of the end point in an optimistic alternative.

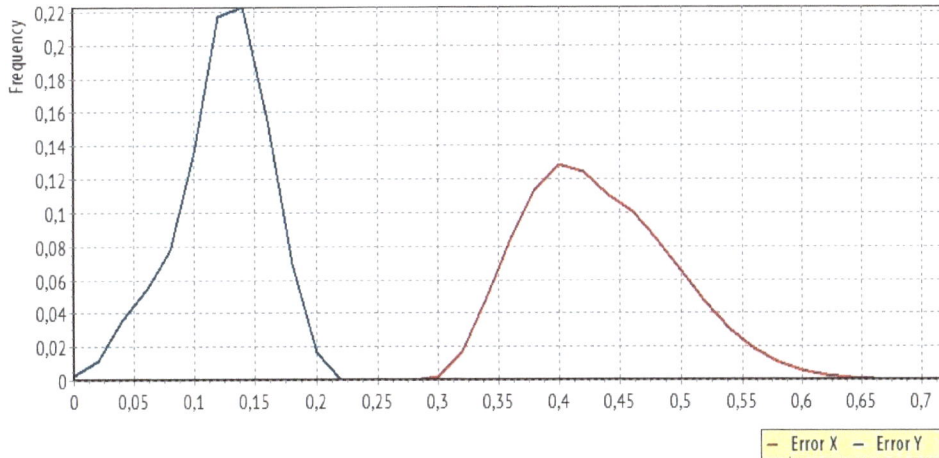

Figure VIII-3 Expected distribution of errors of coordinates X and Y of the end point in a pessimistic alternative.

Figure VIII-2 and Figure VIII-3 show the estimated distributions of the modeled quantities.

The estimated distribution of deviations provides an opportunity for a more reliable assessment of the accuracy of measurement compared to the unit values found by means of "classical" solutions.

VIII.3. ACCURACY OF COMPUTING AREA BY COORDINATES

The formulae for computing the area of a closed polygon by coordinates of its vertices have been known for a very long time but until the 1990s they had no practical significance. After the introduction of geodesic practice of contemporary tools and information technology, however, the determination of areas by coordinates obtained by numerical methods has become the only method applied in practice (M. Mazhdrakov, I. Ivanova, D. Benov, 2014).

By default, the method applies subject to two conditions.

1. The area of a nonintersecting polygon is computed.

2. The coordinates X and Y are determined with the same accuracy.

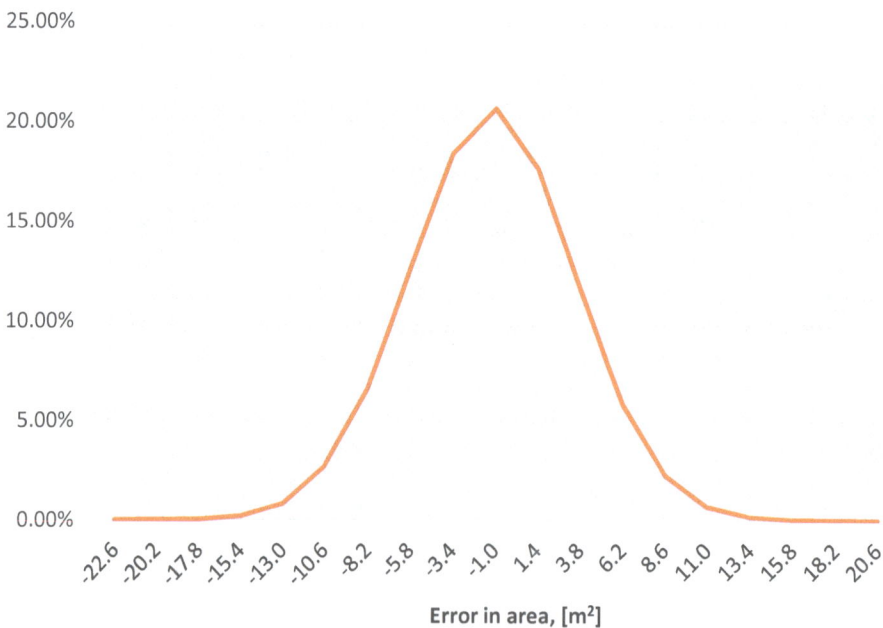

Figure VIII-4. Distribution of error in case of coefficient $k = 1$

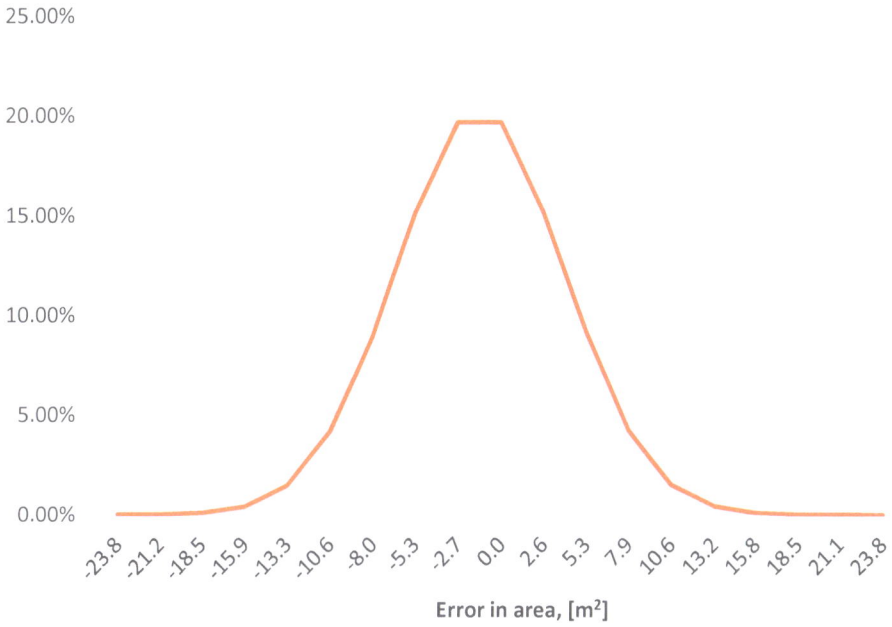

Figure VIII-5. Distribution of error in case of coefficient k = 0.5

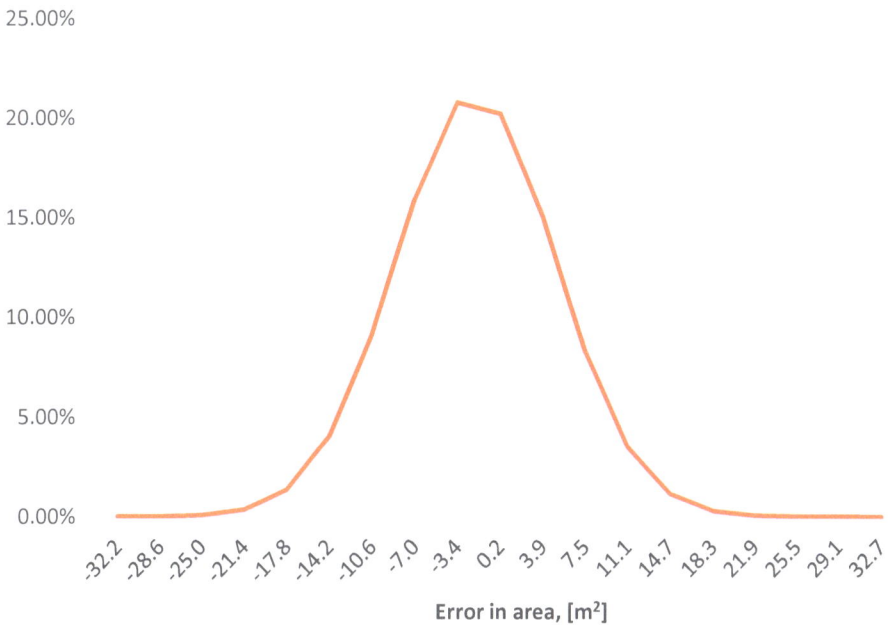

Figure VIII-6. Distribution of error in case of coefficient k = 0.25

When the second condition is not met a systemic error in the area, which depends on figure's orientation and is inadmissible for scalar quantity is assumed.

In geodesic literature the following formula is used to calculate area:

$$P = 0.5 \sum_{1}^{n} X_i(Y_{i+1} - Y_{i-1}), \mathrm{m}^2, \qquad (\mathrm{VIII}.2)$$

or the analogous one –

$$P = 0.5 \sum_{1}^{n} Y_i(X_{i+1} - X_{i-1}), \mathrm{m}^2, \qquad (\mathrm{VIII}.3)$$

where X_i and Y_i are the coordinates of polygon's vertices;

$i = 1, 2, ..., n.$

Formula (VIII.2) and formula (VIII.3), respectively, enable the accuracy assessment to be made by using the Monte Carlo method. Input data are the coordinates of points and their mean squared errors: $X_i \pm m_X, Y_i \pm m_Y$. A normal distribution of errors and achieving the same accuracy along the two coordinate axes is assumed.

Regulated properties in Bulgaria most often have a rectangular shape. Then, if we express the height as a part of the base:

$$H = kB,$$

where $k \leq 1$, we can assess the influence of plot's shape.

Figure VIII-4, Figure VIII-5 and Figure VIII-6 show the results of the application of the Monte Carlo method as regards the area of a rectangular plot of m² and different correlation of the sides at mean squared errors $m_X = m_Y = 0.10\ m.$

Several conclusions can be made from the results obtained. The significance of shape in determining the area by coordinates is confirmed as evident if Figure VIII-4 is compared to Figure VIII-6. In addition, with certain probability (5 to 10%) significant errors can be expected: more than 10 m².

VIII.4.ACCURACY OF COMPUTATION OF VOLUMES IN OPEN-PIT MINES

Open-pit mines are complex extraction systems developing in interaction with constantly changing natural environment. Production machines operate in them of different degree of reliability and capability of adaptation to changed conditions. These, and a number of other factors interfere with the system and, as a rule, the results obtained deviate from the targets set in the management. This necessitates site management with feedback (Figure VIII-7).

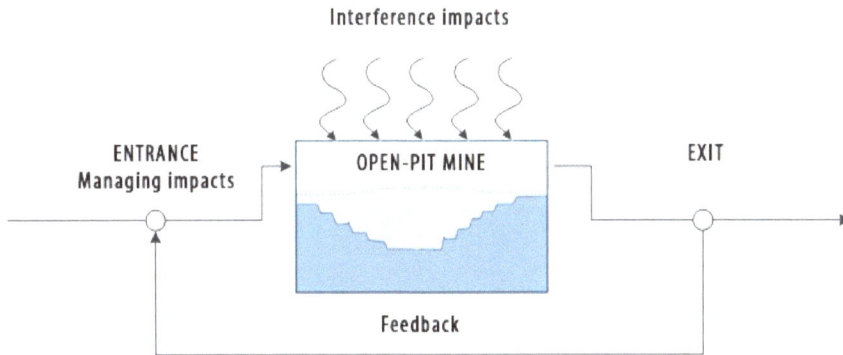

Figure VIII-7. Management with feedback

The term "**feedback**" is fundamental in management theory. Feedback means comparing output information to pre-set (expected) parameters of site's condition. Thus, part of output information is transformed into entry (managing impacts).

In open-pit mines the principal manner of ensuring timely and reliable feedback is to determine the volume of stripped mined mass with a periodic survey work.

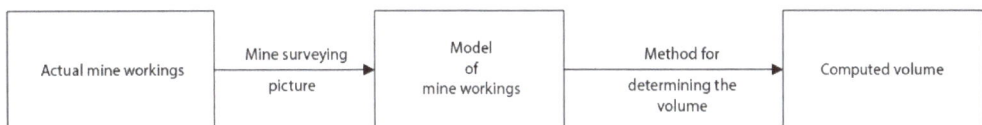

Figure VIII-8. Technology for determination of volume in open-pit mine

Mine workings have quite irregular shape and the stripped mass volume cannot be computed by using the known stereometric formulae. Thus, the volume determination passes through several stages (Figure VIII-8), which is associated with loss of information.

In open-pit mines the volume of irregular object is a function of areas and lengths –

$$V \approx f(P_1, P_2, \ldots P_n, L_1, L_2, \ldots L_m), \qquad \text{(VIII.4)}$$

where areas are designated by P and lengths are designated by L

"Mine surveying" methods of computing volumes differ in terms of position and manner of determination of the said two elements and in terms of the extent to which they approximate the actual shape of stripped volume.

The volume of stripped mined mass is computed most often by the method of vertical parallel sections in equal distance (interval). When determining the volume by the method of vertical parallel sections at the same interval the error in volume is

$$V = L \sum_{i=n}^{K} P_i + VqLP_n + qLP_k, \text{m}^3,$$

where $i = n, n + 1, \ldots, k$ are the numbers of sections between which the volume is determined;

L is the interval between sections, m;

P_i is the area of section i, m^2;

q is the coefficient that takes into account the shape of mined mass outside the terminal sections ($q \leq 1$).

In case of sufficiently small interval $L \leq 10$ m, the volume is

$$V = L \sum_{i=n}^{k} P_i, \text{m}^3, \qquad \text{(VIII.5)}$$

where n and k are the numbers of starting and terminal vertical sections;

P_i is the area of i-th section.

From the formula (VIII.5), by taking into account the shape and the dimensions of sections, the relative error in volume is obtained (M. Mazhdrakov, 1983)

$$\frac{m_V}{V} = \frac{K m_P}{\bar{P}\sqrt{N-1}},$$ (VIII.6)

where $\bar{P}$ is the average stripped area, m²;

m_P is the mean squared error in the area;

K is a coefficient that takes into account the non-uniform distribution of the areas ($K > 1$);

N is the number of sections.

From the formula (VIII.6) it follows that the value of the error $\frac{m_V}{V}$ depends on several groups of indicators:

- mine's productivity, which defines the amount of cross section $\bar{P}$ and the number of sections N;

- the technology of extraction and transportation, on which the area $\bar{P}$, the form of sections and the differences in the areas of neighboring sections (coefficient K) depend;

- the accuracy of survey work on which the mean squared error m_P depends.

Three types of errors are made in tacheometric surveying of the benches of the mine: due to determination of the position of single (detailed) point, due to idealization of the complex shape of mine workings and upon before automated processing: due to graphic concepts.

We will apply the Monte Carlo method for computing of the error in the volume of 3D coordinates of single point and upon idealization of shape by assuming that all other quantities are constant.

Figure VIII-9 shows the expected error in volume due to errors in surveying of the point with a total station $m_X = 0.1$, $m_Y = 0.1$ and $m_H = 0.05$, under normal distribution. It is evident that the error is insignificant, and is normally distributed near zero.

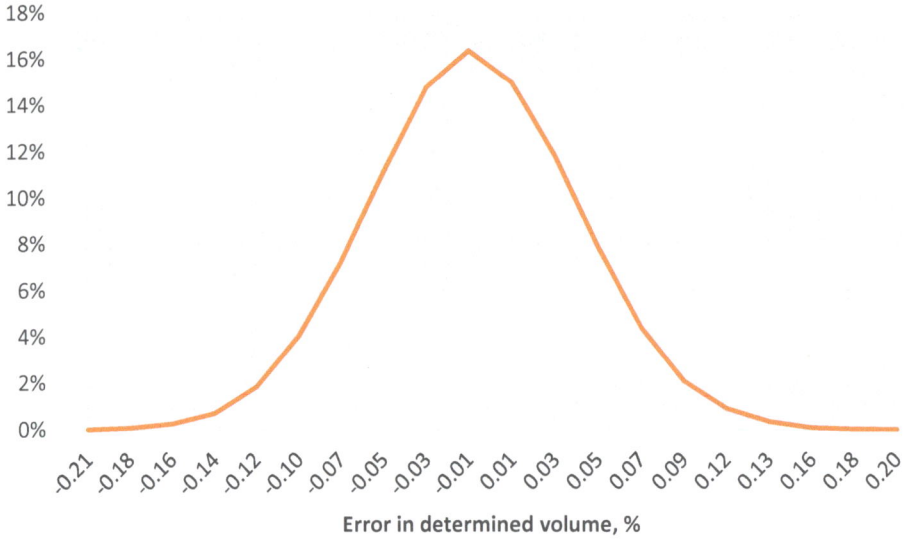

Figure VIII-9. Relative frequency of mean squared error

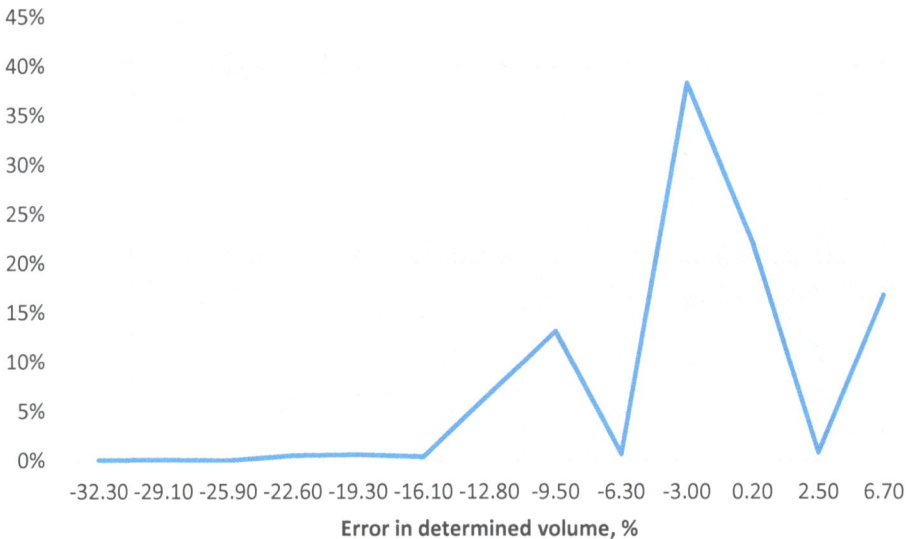

Figure VIII-10. Relative frequency of mean squared error

Figure VIII-10 shows the influence of error due to idealization ("leveling") of the shape of mine workings upon errors $m_X = 1.5$, $m_Y = 1.5$ and $m_H = 0.5$. It is evident that errors are significantly higher, with right asymmetry in the interval from -30 to 7%, as the errors near -3% predominate. This proves once again that the accuracy of surveying depends on the skills of prism carriers and the advantages of remote methods where this error is minimal.

The mean squared error of computed volume depends on the density of distribution of the area of sections: m_p (M. Mazhdrakov, 1983).

Errors in determination of the position of detailed points are due to two sources: to measurement (practically below 0.1 m in present-day measurement devices) and to generalization (idealization) of shape in survey work whereupon prism carriers move on the slopes of the step.

VIII.5. DETERMINING QUARRY DUMP TRUCKS ACTUAL LOAD

For the adequate assessment of efficiency of automobile transport in open-pit mines and hence its optimal management, two questions must be solved (M. Mazhdrakov, D. Benov, G. Trapov, 2013).

1. What is the technical capacity of the skip (container) of the dump truck of a given model?

2. What is the actual load of a specific dump truck?

It is logical to seek the answer to the first question in the main technical data of dump trucks. However, the capacity of the skip is stated there in a wide range. For example, for:

- BELAZ 7513 – with load-carrying capacity of 130 t – from 51 to 74 m³, difference of about 45%;

- BELAZ 7514 – with load-carrying capacity of 120 t – from 47 to 61 m³, difference of about 30%.

This difference is not accidental. It reflects the fact that in bulk non-homogeneous cargo the volume of cargo is not a technical but a technological indicator. Thus, under the defined conditions the answer to the first question must be searched by solving the second question, i.e. by measuring the actual load of the dump truck.

The single-image photogrammetry is suitable for determining the volume of the loaded material.

The method has some substantial advantages:

- the actual (expanded) volume of loaded mined mass is measured.

- it also makes it possible to determine other characteristics: granularity, angles of load's inclination, etc. and by using them: the shape that best suits the geometric parameters of the skip and type of the cargo;

- they use accessible architecture and simple technologies.

The use of digital photo camera enabled the development of a technology with considerable volume of automated operations.

In order to apply single-image photogrammetry certain conditions must be met:

- the shooting axis must be horizontal and aimed perpendicular at the object;

- the image must be scaled by linear details of known length;

- if the focal length f of the camera and the image enlargement (the "positive") v are unknown the distance to the object should be measured.

A photograph of BELAZ 130 made as per the above conditions is given in Figure VIII-11 and thereon a coordinate system zOx is built.

The photograph is scaled by using two intercepts: a_1 (horizontal) and a_2 (vertical). The scales are:

$$M_h = \frac{A_1}{a_1} \text{ and } M_v = \frac{A_2}{a_2},$$

where A_1 and A_2 are the actual lengths.

It is necessary that $M_h \approx M_v$.

The graphic coordinates x_i and z_i, $i = 1,2, \dots 7$ are measured. All measured quantities must be in the same unit of measure.

For the accuracy assessment we apply the Monte Carlo method. From the said algorithm it follows that the error in volume is a function of the errors in measured quantities on the photograph

- $m_{a1} = m_{a2}$;
- The graphic coordinates of points 1, 2, 3, 4, 5 and 6;
- the interpolated Z of 8, 9 and 10.

Figure VIII-11. Photography with inserted coordinate system

The spatial coordinates of the points are computed:

- for $i = 1,2,3$ (distance S): $X_i = M_h x_i$, $Z_i = M_v z_i$;
- for $i = 4,5,6$ (distance $S + B/2$): $X_i = (1 + 0.5\,B/_S)M_h x_i$, $Z_i = (1 + 0.5\,B/_S)M_v z_i$;
- for $i = 7$: $X_7 = X_1$, $Z_7 = (1 + 0.5\,B/_S)M_v z_7$;

The coordinates X_i of invisible points ($i = 8,9,10$) are $X_8 = X_4, X_9 = X_5$ and $X_{10} = X_6$;

The coordinates $Z_{8,9,10}$ are interpolated.

Table VIII-1. Coordinates of characteristic points registered by the photograph.

No.	X	Y
1	5.223	1.743
2	6.301	3.648
3	6.064	6.219
4	4.561	7.545
5	3.814	3.814

Figure VIII-12. Results f simulation under Monte Carlo method

The volume of the skip is

$$V_k = P_{123}B, \mathrm{m}^3,$$

where P_{123} is the area of the triangle 123, m²;

B is the width of truck's skip, m.

The simulation under the Monte Carlo method has been made by using the coordinates given in Table VII-1, and at width of truck's skip of $B = 5.20$ m. The Guesstimate software has been used. The results of the simulation are shown in Figure VIII-12.

VIII.6. ERROR IN MEASURING AN ANGLE DUE TO ECCENRICITY OF THEODOLITE AND SIGNALS

Constructive solutions of new technical means of measuring angles and lengths limit to the minimum the so-called personal errors at sighting and reading. Thus the impact of other errors has increased, such as the errors due to the deviation of points (eccentricity) of the theodolite and signals.

The measurement with three tripods and a forced-centering kit reduces but does not exclude the influence of this error because differences between the center of the point and the vertical axis of the tool (signal) are presumed.

The analysis of eccentricity's impact is one of great importance because it is underlying for he practical rules on angle measurements. The classical conclusion of formulae is complex (K. Velev, 1965; I. Ushakov, D. Kazakovskii, G. Krotov et al., 1989); the shorter conclusion (M. Mazhdrakov, I. Ivanova, 2014) refers to the special cases where the eccentricity's impact is the strongest.

The influence of eccentricity is illustrated in Figure VIII-13. The centers of the points are A, B and C, and the vertical axes of the theodolite and signals are A', B' and C'. The position of these axes is determined by the length e and the pointing angle α of deviations (eccentricity) AA', BB' and CC'. The difference between the angle measured and the actual angle is

$$\Delta\beta = \measuredangle ABC - \measuredangle A'B'C', \tag{VIII.7}$$

and it can be expressed by the difference between the pointing angles computed by using the coordinates of the points and A', B', C'.

From Figure VIII-13 it is evident that at constant parameters: angle β and side length S, the coordinates of points A', B' and C' depend on the values of e and α; e.g. for point A' –

$$\left.\begin{aligned} X'_A &= X_A + e\cos\alpha, \\ Y'_A &= Y_A + e\sin\alpha, \end{aligned}\right\} \tag{VIII.8}$$

In modeling we assume that the linear deviation e changes by the normal law and the pointing angle α changes uniformly from 0 to 400 gon.

The angle β and the lengths between the points are computed by the formulae of the second main problem.

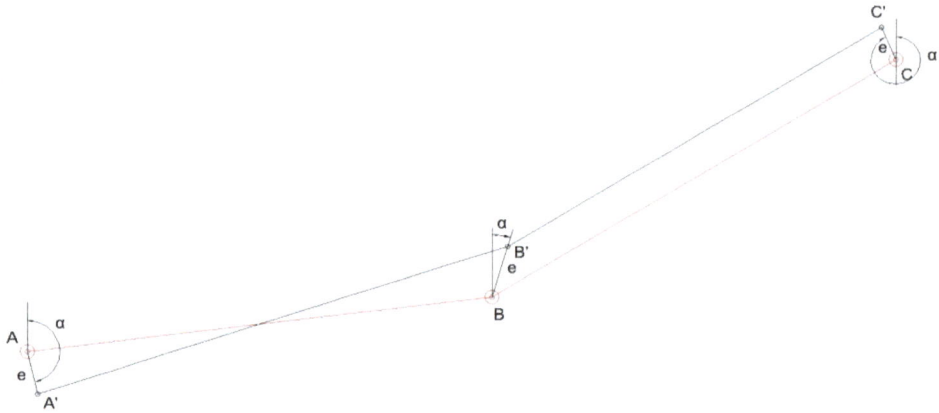

Figure VIII-13. Influence of eccentricity

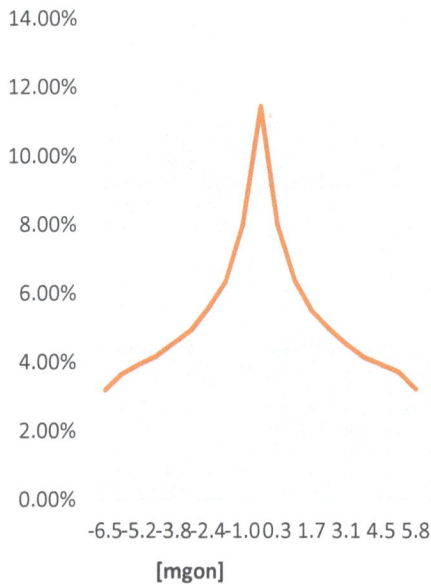

[mgon]

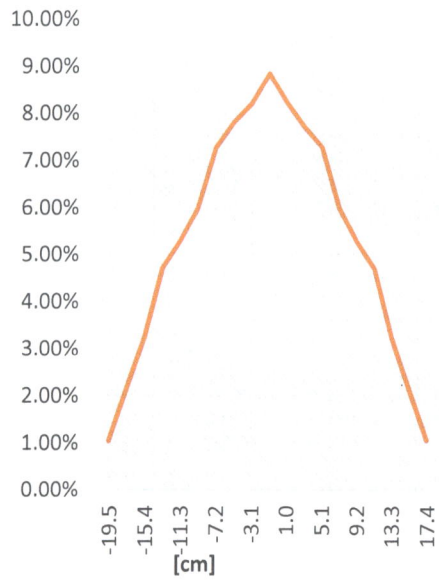

[cm]

Figure VIII-14. Influence of eccentricity when measuring a small angle: 1 gon, at same length of sides

Figure VIII-15. Influence of eccentricity when measuring a small angle: 1 gon, in case of side ration of 1:3

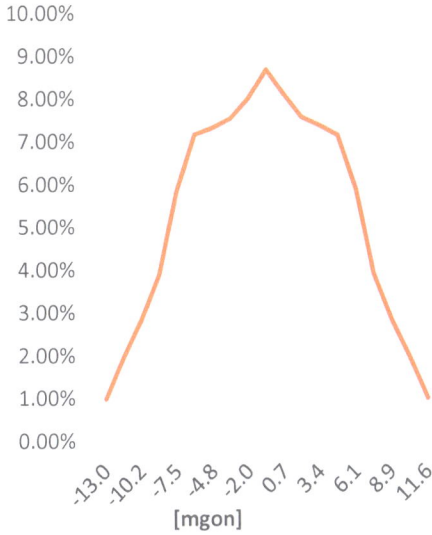

Figure VIVIII-16. Influence of eccentricity when measuring a flat angle: 199 gon, , at same length of sides

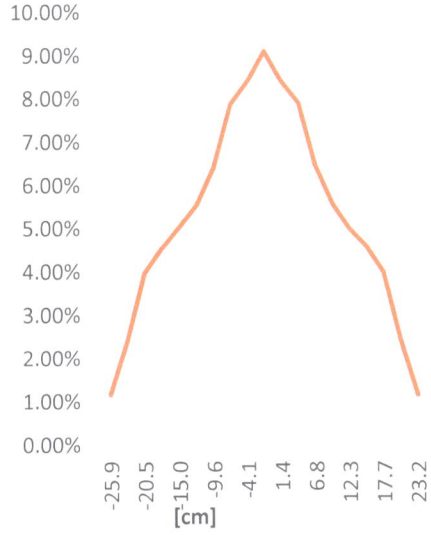

Figure VIVIII-17. Influence of eccentricity when measuring a flat angle: 199 gon, in case of side ration of 1:3

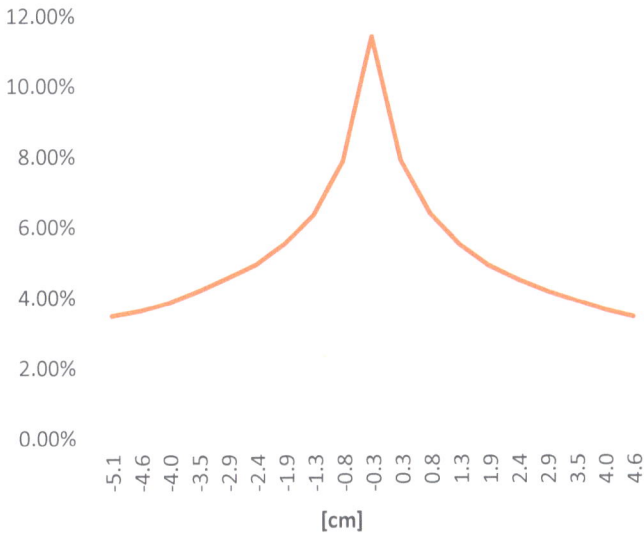

Figure VIII-18. Absolute error in measured distance

In order to assess the influence of the main factors we model the error in 4 variants (Table VIII-2) at $e = 5$ mm and mean squared error $m_e = \pm 2$ mm. The length of the side is 500 m.m.

Table VIII-2. Options for error modeling

Angle	Ratio of the lengths of sides	
Acute (1 gon)	1:1 (Figure VIII-14)	0.33:1 (Figure VIII-15)
Flat (199 gon)	1:1 (Figure VIVIII-16)	0.33:1 (Figure VIVIII-17)

The results confirm the influence of angle's size on the error due to eccentricity of the tool or the signals. The ratio of sides' lengths does not have crucial significance for the amount of the error but it increases the probability of occurrence of errors of greater amount.

REFERENCES

[1] Bahurin, I. (1936). <u>Questions of mine surveying</u>. Moscow-Leningrad [in Russian].

[2] Cheshankov, L., P. Ganchev, S. Harizanov, et al. (1969). <u>Technical Mine Surveying Instruction</u>. Sofia, Tehnika [in Bulgarian].

[3] Mazhdrakov, M. (1983). <u>Mine Surveying</u>. Sofia, VMGI [in Bulgarian].

[4] Mazhdrakov, M., D. Benov, I. Ivanov (2012a) Forecasting the accuracy of "open" polygon with stochastic modeling (method "Monte Carlo"). 4.

[5] Mazhdrakov, M., D. Benov, I. Ivanov (2012b). <u>Predicting the Mine Surveying Measurement Error with the Monte Carlo Method</u>. Third National Scientific and Technical Conference with International Participation "Technologies and Practices in Underground Mining and Mine Construction", Devin [in Bulgarian].

[6] Mazhdrakov, M., D. Benov, G. Trapov (2013). <u>CADMin. Automated Planning of Development of Mining Works in Open Pits</u>. Sofia, MGU St. Ivan Rilski [in Bulgarian].

[7] Mazhdrakov, M., I. Ivanova (2014). <u>Geodesy Part I</u>. Shumen, Bulgaria, Konstantin Preslavsky University of Shumen [in Bulgarian].

[8] Mazhdrakov, M., I. Ivanova, D. Benov (2014). Accuracy of calculating area by coordinates. <u>Geomedia</u>. Sofia. 2: 36-37 [in Bulgarian].

[9] Mazhdrakov, M., T. Trendafilov (1981). "Application of statistical modeling (Monte Carlo method) for the investigation of underground polygon errors." <u>Rudodobiv</u> 6: 13-14 [in Bulgarian].

[10] Ushakov, I., D. Kazakovskii, G. Krotov, et al. (1989). <u>Mine Surveying</u>. Moscow, Nedra [in Russian].

[11] Velev, K. (1965). <u>General Mine Surveying with Analysis of Mine Surveying Works</u>. Sofia, Tehnika [in Bulgarian].

SECTION IX.
EMISSIONS MODELING

V. Hristov, St. Topalov, M. Mazhdrakov, D. Benov

O ver the past years, scanning and navigation systems as well as express analyses ensuring a current flow of information have occupied a major place in environmental protection monitoring. On the basis of express analysis it became possible to develop real-time forecasting systems. Modeling by the Monte Carlo method would play an important role here.

Environment monitoring systems play a major part: presence of gases and dust in the air, of various pollutants in water basins and subsoil waters, etc. The early warning systems fed with data from the monitoring need adequate models of the areas of pollution (J. R. Stokes, A. Horvath, 2009).

In modeling of pollution by different and/or organic pollutants of the atmosphere, respectively water basins, the following indicators must be determined:

- pollutant's type;
- pollutant type: point-source or diffuse;
- duration of the effect;
- the area affected by the pollutant.

Monte Carlo modeling would also play a role.

IX.1. STUDYING AEROSOLE POLLUTION

Pasquill and Gifford's analytical model for dispersion of aerosol concentration in the atmospheric air (V. Hristov, 2013) is:

$$C = \frac{Q}{\pi u \sigma_x \sigma_y} e^{-\frac{1}{2}\left[\left(\frac{H}{\sigma_x}\right)^2 + \left(\frac{y}{\bar{\omega}_y}\right)^2\right]}, \qquad (IX.1)$$

where C is the concentration of aerosol computed on the level of the earth at certain distance from the source;

Q is the quantity of emission of the source, g/s;

u is the average wind speed, m/s;

y is the distance from source in the coordinate system (x, y) having as its center the source of pollution and the axis x is windward;

H is the effective height of the chimney, m;

σ_x and σ_y are the coefficients of aerosol diffusion by the axes x and y.

Dependency (IX.1) provides actual values of the concentration C as a function of x and y only in the semiplane defined by wind direction. Determination of the diffusion coefficients σ_x and σ_y is carried out on the basis of an experimentally taken family of curves setting their dependence on the distance x from the source for six averaged types of atmospheric conditions. The six types correspond to wind speeds between 0 and 2, 2 and 6 and greater than 6 m/s and to strong, weak and low solar radiation. For the analysis a weak instable type of atmospheric conditions has been chosen (u is 4 to 6 m/s and at average solar radiation), which corresponds to the most common type of atmospheric conditions.

The coefficients σ_x and σ_y are determined on the basis of linear regression models:

$$\sigma_x = a_1 x + b_1, \sigma_y = a_2 x + b_2, \qquad (IX.2)$$

where for coefficients a_1 a_2, b_1 and b_2 the values 0.58, -0.97, 0.028 and 55.01 have been obtained.

For the remaining parameters the following values are assumed:

— height of source H= 80 m;

— wind speed u=6 m/s, which corresponds to a weakly instable type of atmospheric conditions;

— quantity of emission Q = 579.05 g/s, corresponding to the emission of Kremikovtsi.[21] Thermal Power Plant.

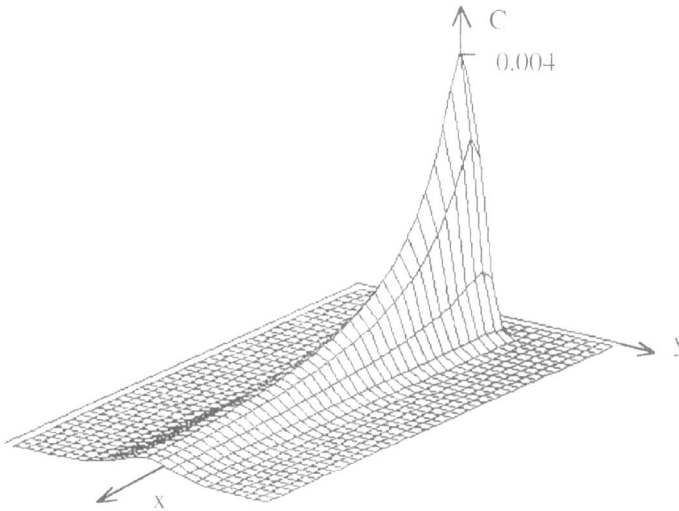

Figure IX-1. Change in concentration C as a function of x and y

The graph of change in concentration C as a function of x and y (Figure IX-1) is obtained by applying a dependency (IX.1) to the assumed parameters. On the graph it is evident that the highest concentration is at the place of the source and that concentration gradually decreases exponentially in the direction x of the wind while in the direction y, which is perpendicular to wind direction the concentration is close to the normal diffusion.

The suggested model can also be applied for analysis of pollution with gases and airborne dust (with particle size up to 50 µк). In this case, however, other diffusion coefficients must be

---

21 Kremikovtsi is an industrial district of Sofia, Bulgaria.

selected. The model has practical application for a distance up to 100 km from the source of pollution.

By the Monte Carlo method the process has been modeled for a distance along the axis x of 200, 400, 950 and 1 200 m. In case of varying indicators Q (450; 550) and u (4.5; 7.5), at chimney height H (60, 120) and H (75, 95). The graphs of distribution of the expected concentration are shown in Figure IX-2 and Figure IX-3.

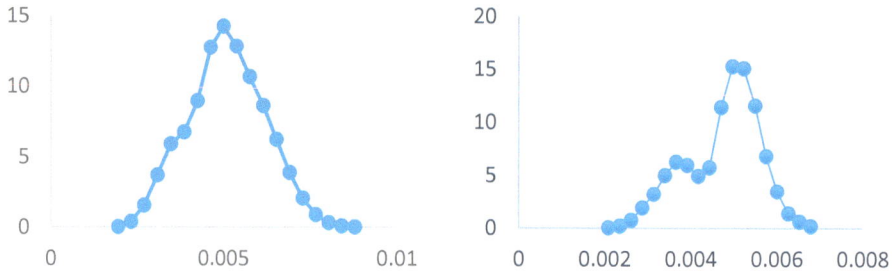

Figure IX-2. Expected concentration at a distance of 200 m and chimney height H = 60 ÷ 120 m (left) and at H = 75 ÷ 95 m (right)

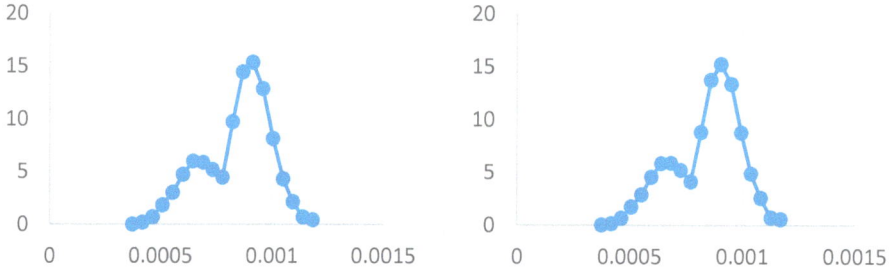

Figure IX-3. . Expected concentration at a distance of 950 m and chimney height H = 60 ÷ 120 m (left) and at H = 75 ÷ 95 m (right)

IX.2. POLLUTANTS IN WATER BASINS

The dispersion of concentration of point-source pollutant in a water basin is examined in case of a one-off (salvo) impact and in case of continuous impact (for certain period of time). The dispersion of the concentration of a substance dissolved in a liquid with mass m for the time of dispersion t and distance from the point of discharge r, is determined by the dependency (V. Hristov, 2013)

$$C(r,t) = \frac{m}{8\rho(\pi D t)^{\frac{3}{2}}} e^{-\frac{r^2}{4Dt}}, \qquad (IX.3)$$

where D is the diffusion coefficient of the pollutant;

ρ is the density of the liquid.

From the theory of liquids it is known that the diffusion coefficient for homogeneous liquid changes by the law:

$$D = \frac{d^3}{6\tau_0} e^{-\frac{M}{kT}}, \qquad (IX.4)$$

where d is the average distance between the molecules of the defined type of liquid;

τ_0 is the average period of variation of molecules around the state of equilibrium;

W is the energy of activation of molecules of the respective liquid;

T is the temperature.

From (IX.4) it follows that the diffusion coefficient increases by increasing the temperature. This fact is explained by the increase of the distance between the molecules of the liquid and the decrease in time for relaxation (the period of variation) of the molecules in it.

Under real conditions, the process of dilution of waste water in river basin must be taken into account. In this process the following physical phenomena can be distinguished:

– dilution in homogeneous environment;

– dilution in stratified environment;

– initial dilution near the source of discharge;

– secondary dilution when the coefficients of horizontal and vertical diffusion are taken into account;

– process of dilution in round and flat jet.

The degree of initial dilution N_0 can be defined by the Cedervall equation (V. Hristov, 2013):

$$N_0 = 0.54 Fr_0 \left(0.38 \frac{z_0}{d_0 Fr_0} + 0.66 \right)^{\frac{5}{3}} \qquad \text{(IX.5)}$$

where Fr_0 is the general Froude number;

z_0 is the depth of discharge of pollution source;

v_0 is the diameter of discharge outlet;

d_0 is the speed of the jet leakage.

Froude number is determined by the formula:

$$Fr_0 = \frac{v_0}{\sqrt{\dfrac{\rho_s - \rho_0}{\rho_0} g d_0}}. \qquad \text{(IX.6)}$$

where the coefficients ρ_s and ρ_0 are the densities of waste liquid and water from the basin. To avoid the phenomenon of intrusion (reverse infusion of water from the basin into the discharger) it is necessary that $Fr_0 > 1$. The extent of secondary dilution can be determined by the method of T. Gardanov for stationary sources and variable coefficient of horizontal turbulent diffusion:

$$N = \frac{\left(B_0^{2/3} - 0.08 \dfrac{x}{v} \right)^{3/2}}{B_0} = \frac{A}{B_0}, \qquad \text{(IX.7)}$$

where A is the width of diffusing field of pollutants;

B_0 is the length of the diffusing discharger, v is the average speed of river stream;

x is the distance from the point of discharge.

The coefficient of horizontal turbulent diffusion can be obtained by:

$$D_x = 0.01 \left(B_0^{2/3} + 0.08 \frac{x}{v} \right)^2. \qquad \text{(IX.8)}$$

The graph of change in the coefficient of horizontal diffusion as a function of x is given in Figure IX-4. From the graph it is evident that this coefficient increases as a parabola by increasing the distance from the source x and for a distance of 10K m it reaches a value of 1600.

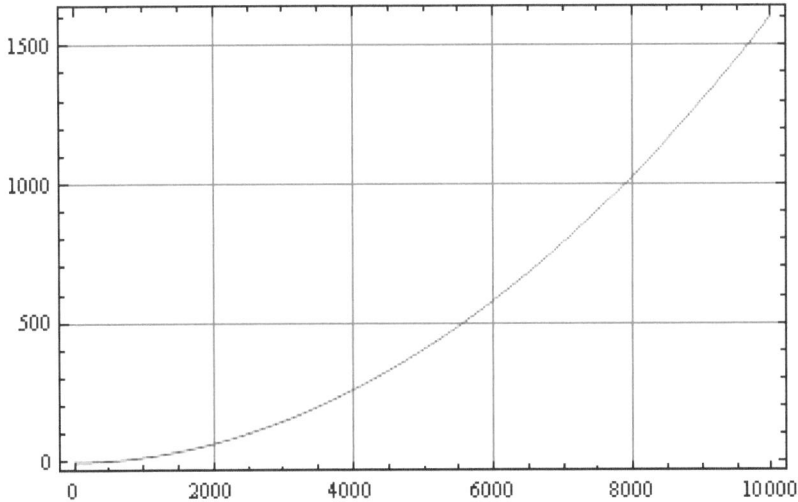

Figure IX-4. Change in coefficient of horizontal turbulent diffusion D_0 as a function of x

Based on the relations on the dispersion of the concentration of substance dissolved in liquid (IX.3), the following cases can be examined.

1. *One-off discharge of pollutant into a water basin with no stream.* The dispersion of concentration C in a given point (x, y) of water surface at the moment t is obtained after replacing the distance r from the center of the coordinate system by $r^2 = x^2 + y^2$ in the formula (IX.3)

$$C(, x, y, t) = \frac{m}{8\rho(\pi Dt)^{\frac{3}{2}}} e^{-\frac{x^2+y^2}{4Dt}}, \qquad (IX.9)$$

where m is the mass of pollutant;

ρ is water density;

D is the diffusion of pollutant in the water.

The start of the coordinate system is the point of infusion of the pollutant.

The concentration spreads circularly around the point of pollution as the highest concentration is in the center and the farther one goes the more it decreases (Figure IX-5). Over time the polluted spot increases but the concentration decreases (Figure IX-6).

In this case the problems as regards the determination of the following values can be posed:

-   the diameter or the area of pollution greater than the set concentration at a given moment in time;

-   in how much time will the pollution at the point of discharge drop below some set maximum concentration limits C_{MCL}.

These problems are solved in the following manner:

In the first problem the diameter is $\left(\frac{d}{2}\right)^2 = x^2 + y^2$. When it is replaced in (IX.9), and the resulting problem is solved with respect to d, one gets

$$d(t, C) = 4\sqrt{Dt \log \frac{8C\rho(\pi Dt)^{\frac{3}{2}}}{m}}. \qquad \text{(IX.10)}$$

In the second problem $x^2 + y^2 = 0$ and from (IX.9) we find t.

$$t = \frac{1}{4\pi D}\left(\frac{m}{\rho C_{MCL}}\right)^{\frac{2}{3}}. \qquad \text{(IX.11)}$$

2. One-off discharge of pollutant into a water basin with a stream. We examine (IX.9) as an equation in relation to x. After that equation is solved we get the distance along the axis x traveled by the pollutant of a given concentration C at a given point in time t. If we add to this distance vt, i.e. the additional distance traveled by the pollutant as a result of the stream the concentration C, will be

$$C(, x, y, t) = \frac{m}{8\rho(\pi Dt)^{\frac{3}{2}}} e^{-\frac{(x-vt)^2+y^2}{4Dt}}, \qquad \text{(IX.12)}$$

where v is the speed of the stream.

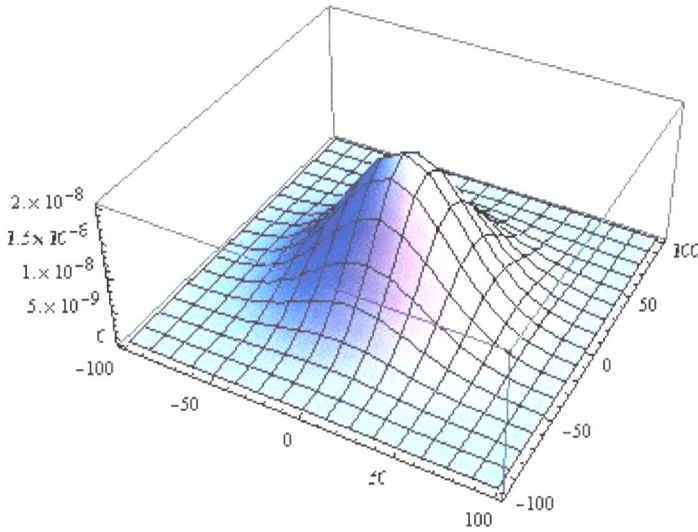

Figure IX-5. Dispersion of the concentration in a basin with no stream in case of one-off discharge of pollutant

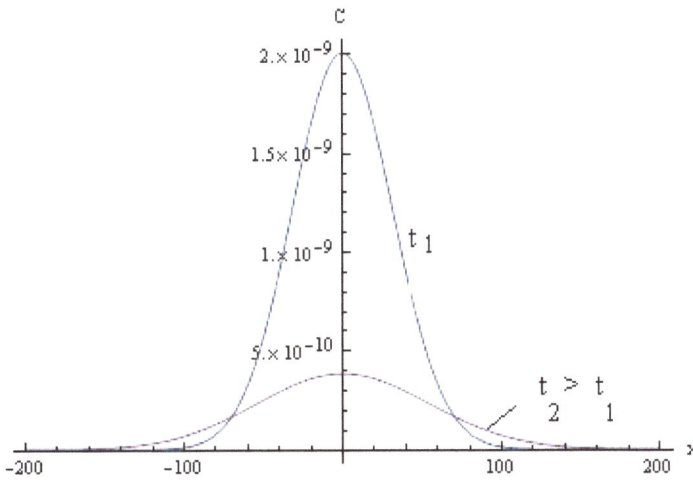

Figure IX-6. Dispersion of pollution in moment t_1 and moment t_2

The coordinate system is so oriented that the axis x coincides with stream direction. The spot with the pollutant is again in the form of a circle that widens by the passage of time and the concentration decreases. In contrast to the preceding case this circle is carried away downstream (Figure IX-7).

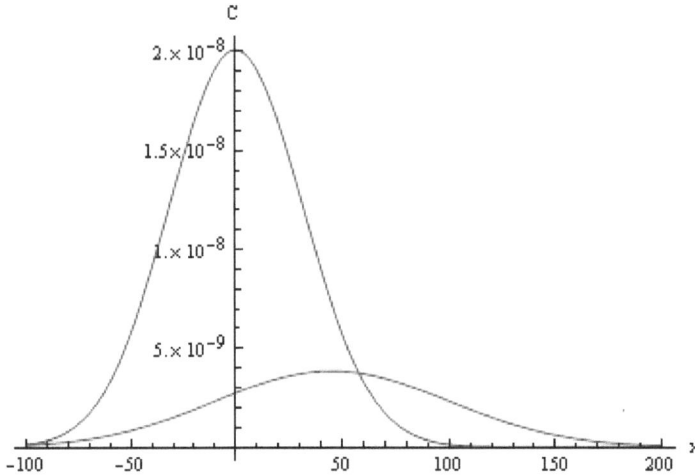

Figure IX-7. Spread of pollutant in a basin with stream in case of one-off discharge

The described model is the basis for the solution to the following problems.

1. Determining the moment t, when a pollutant with maximum concentration passes through a given point at the water basin.

The first derivative is looked for by the time of $C(, x, y, t)$ from (IX.12). The obtained expression is equaled to 0 –

$$\frac{Dm(-6Dt - t^2v^2 + x^2 + y^2)}{32\pi^{\frac{3}{2}}\rho(Dt)^{\frac{7}{2}}e^{\frac{(x-vt)^2+y^2}{4Dt}}} = 0. \qquad (IX.13)$$

In order to have a solution the equation (IX.13) must $-6Dt - t^2v^2 + x^2 + y^2 = 0$.

This is a quadratic equation that has a solution:

$$t = -\frac{3D + \sqrt{9D^2 + v^2(x^2 + y^2)}}{v^2}. \qquad (IX.14)$$

1. Finding the location of an unknown source of pollution, time and quantity of discharged pollutant.

We have to find three points from the spot of the same concentration C_1 the centre P and the radius R of the circle passing through the three points. The concentration C_P is measured in the point P, and then by the formula

$$t = \frac{R^2}{4D \lg \frac{C_P}{C_1}},$$ (IX.15)

and

$$m = 8C_P\rho(\pi D t)^{\frac{3}{2}}$$ (IX.16)

the unknown mass m and time t of pollution are determined and the source is found at a distance vt from point P upstream.

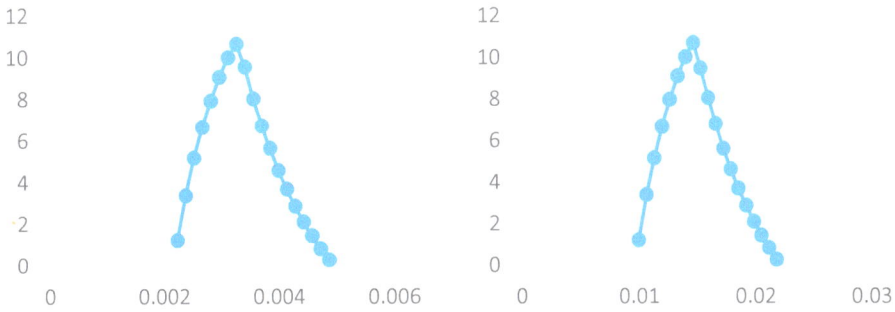

Figure IX-8. Results of the run at a distance of 50 m (left) and 150 m (right)

From Figure IX-8 it is evident that under the specified conditions no significant deviations of modeled parameters can be observed.

REFERENCES

[1] Hristov, V. (2013). Modeling of indicators of deposits of underground resources and related objects through computer systems. Mine Surveying. Sofia, MGU St. Ivan Rilski. Doctor [in Bulgarian].

[2] Stokes, J. R., A. Horvath (2009). "Energy and Air Emission Effects of Water Supply." Environ. Sci. Technol. 43(8): 2680-2687.

SECTION X.
URBAN ACOUSTICS

N. Nikolov, D. Benov, M. Mazhdrakov, D. Benova, E. Mitova

Contemporary bog city (urban agglomeration) is the noisiest place for work and/or dwelling chosen by a considerable part of mankind. We should also note that noise is (1) extremely aggressive and widespread form of pollution of urbanized environment, and (2) the growth of urban population and their mobility lead to an increase in noise, including the emergence of new sources.

Most researchers put in the first place the impact of traffic noise. This is determined by the increasingly developed street network, increase in number of automobiles, irrational traffic organization and other suchlike factors. However, there are some other noise sources, some of them related to the specific nature of urban life, that should not be underestimated (N. Nikolov, A. Kovachev, D. Benov et al., in print).

X.1. NOISE EMITTED BY TRAFFIC FLOW

Noise level of the source is a fundamental parameters in acoustic computations. The noise level can be obtained by using three main methods.

1. Analytically, by using equations that take into account the relations between the noise level and objectively identifiable

quantities. Usually, these formulae are regulated in statutory documents.

2. By observation of the actual state of noise conditions (natural or field measurements).

3. From a noise map.

The analytical method is fundamental in designing new roads or in reconstruction of existing ones.

The specialized literature contains a considerable number of formulae for computing the noise level of different sources, in the first place – of automobile flows. The formulae are obtained by means of a similar methodology with appropriate processing (multi-factor analysis) of data from acoustic measurements. For this reason the formulae reflect the actual state of the vehicle fleet, of road network, and, to an extent, of the specific features and behavior of traffic participants.

The comparison of published formulae makes it possible to draw some conclusions. The indicators of traffic flow, such as traffic intensity and speed and the structure of the flow (participation of one type of vehicles or another) have fundamental importance for the value of the noise level. The type of road pavement and road slope, the number and width of traffic lanes, etc. are also reflected.

The great influence of "national" peculiarities makes it difficult to prepare a common formula for determining the noise level. Thus, national statutory documents regulate different formulae. These are work formulae of similar stepped structure meeting the traditional computing means of mid-XXth century.

The common principles for calculating the noise level are expressed by means of the equation –

$$L_{(A),(eq,max,...)} = E - D_d - D_a - D_g - D_b - C_r - -C_m, \text{dB or dB(A)}, \quad \text{(X.1)}$$

where $L_{(A),(eq,max,...)}$ is the noise level at the point of computation;

E is the level of noise emission or the level of noise emitted from the source at the reference point;

D_d is the decrease in the noise level depending on the distance between the source and the point of computation;

D_a is the decrease in noise due to atmospheric conditions;

D_g is the decrease in the noise due to earth surface absorption;

D_b is the decrease due to noise barrier facilities;

C_r is the adjustment for reflection;

C_m is the adjustment for meteorological conditions.

Table X-1. Correction taking into account the longitudinal slope depending on the structure of traffic flow

Longitudinal slope, %	ΔL_{Asl}, dBA Structure of the transport stream (% freight vehicles)				
	0	5	20	40	100
2	0.5	1	1	1.5	1.5
4	1	1.5	2.5	2.5	3
6	1	2.5	3.5	4	5
8	1.5	3.5	4.5	5.5	6.5
10	2	4.5	6	7	8

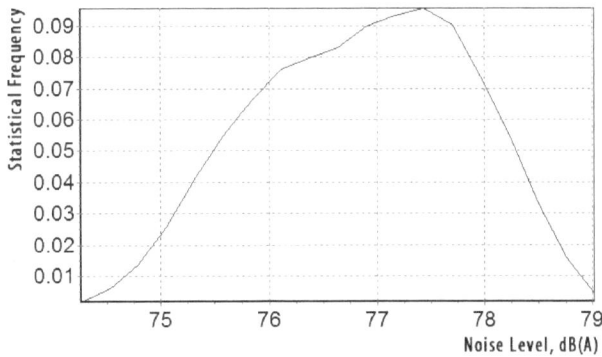

Figure X-1. Results from the application of the Monte Carlo method

On the basis of a deterministic mathematical model of traffic flows the following equation has been derived (G. Osipov, 2004):

$$L_{Aeq}(7.5) = 10\lg N + 13.3\lg V + 4\lg(1 + \rho) +$$
$$+\Delta L_{Asurf} + \Delta L_{Asl} + 15, dB(A), \tag{X.2}$$

where N is two-way traffic intensity, veh/h;

V is the average traffic speed, km/h;

ρ is the relative share of trucks in the flow, %;

ΔL_{Asurf} is an correction taking into account the type of road paving, dB(A) (if asphalt concrete and asphalt – $\Delta L_{Asurf} = 0$, if concrete or rough asphalt concrete $\Delta L_{Asurf} = +3$ dB(A));

ΔL_{Asl} is an correction taking into account the influence of longitudinal slope depending on the structure of traffic flow, dB(A), which is reported as per Table X-1.

The constant parameters are the adjustments for the technical qualities of automobiles, existence of dividing line, longitudinal slope above 2%, two-side or one-side building development

Variable indicators in this case are the traffic flow intensity, speed of motorcars, share of trucks.

Model results of the application of the method are given in Figure X-1.

X.2. PROPAGATION OF QUASI-CYLINDRICAL SOUND WAVES

In acoustics, two types of propagation of sound waves are studied: from point source or from linear source. In the first case the noise energy propagates in the form of a sphere, and in the second one – as a cylinder of endless length (L. D. Landau, E. M. Lifshitz, 1987).

It is accepted that the decrease in noise level is computed at a double increase of distance. Sound energy decreases proportionately to the front, i.e. of r^2 or of r, where r is the distance from the noise source to the point of computation. Then, for the spherical wave the decrease will be

$$\frac{\partial^2 p}{\partial t^2} = c_0^2 \left(\frac{\partial^2 p}{\partial r^2} + \frac{2}{r} \frac{\partial p}{\partial r} \right). \qquad (X.3)$$

and for the cylindrical one –

$$\frac{1}{r} \frac{\partial}{\partial r} \left(r \frac{\partial \varphi}{\partial r} \right) - \frac{1}{c^2} \frac{\partial^2 \varphi}{\partial t} = 0. \qquad (X.4)$$

From formulae (X.3) and (X.4) it follow that the decrease in noise level when increasing the distance has the same gradient that do not depend on the distance between the source and the point of calculation and in case of doubling it can take values of only 3 or 6 dB(A).

This conclusion contradicts the nature measurements in actual urban environment and the results of laboratory experiments. Therefore, in 1985 (N. Nikolov, D. M. Benov, M. Mazhdrakov) defined a new class of sound waves defined by him as "quasi-cylindrical".

In terms of physics, quasi-cylindrical waves are generated by arranged point sources emitting cophasially [sinfazno]. For this reason it can be expected that in case of a limited number of sources the waves will have properties close to the spherical waves and in case of unlimited number the properties will be close to the cylindrical ones. Therefore, if the distance is doubled, the decrease in noise will be not less than 3 dB(A), and the increase will be not greater than 6 dB(A).

Differential equation of the propagation of quasi-cylindrical sound waves looks like this

$$\frac{\partial^2 p(r,t)}{\partial t^2} = c_0^2 \left\{ \frac{1}{r^n} \left[\frac{\partial}{\partial r} r^n \frac{\partial p(r,t)}{\partial r} \right] \right\}, \qquad (X.5)$$

As a physical model the source of quasi-cylindrical waves with unlimited number of emitters is closest to traffic flow in highways and thoroughfares. The flow is characterized by the distance occupied by one automobile (its own length + distance): ℓ, m. Then, if the distance is doubled the noise level decreases by

$$\Delta L(r) = 10 \left(1 + \frac{\lg \frac{\ell}{\pi}}{\lg r} \right) \lg \frac{r}{7.5}. \qquad (X.6)$$

When $\frac{\lg \frac{\ell}{\pi}}{\lg r} > 1$, the point of calculation is within the area of priority propagation of spherical sound waves emitted by a single point source. This area spreads at a distance of $r < \frac{\ell}{\pi}$. When $r = \frac{\ell}{\pi}$, $\Delta L(r) = 20 \lg \frac{r}{7.5}$.

Based on a formula (X.6) a Monte Carlo procedure has been applied for the propagation of quasi-cylindrical sound waves.

The main input quantities are:

- the interval between single sources, m;
- the distance (30, 60, 120 and 240 m).

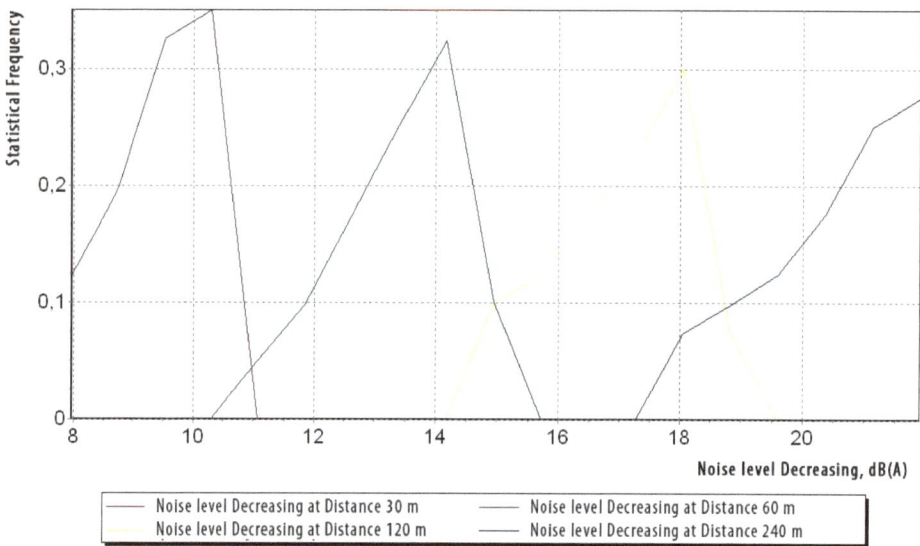

Figure X-2. Decrease in noise level depending on the distance from source

From the results obtained (Figure X-2) it can be concluded that in case of high intensity – an interval between automobiles below 50 m – a manifested delta distribution is established as in case of lower intensity it is more is less manifested. In all cases the distance between the maximums is about 4 dB(A).

X.3. DETAILED MODELING OF THE NOISE LEVEL OF TRAFFIC FLOW

The noise level of the source is the level of emitted sound measured under certain conditions: distance, frequency, terrain height and others.

The noise level of traffic flows, and of traffic flows along the highways in particular, is determined most often in two ways:

- analytically, as a function of certain number of parameters, and

- by measuring the noise level that could be done immediately (on the terrain) or on a noise map.

The analytical method is fundamental in designing new sites or in reconstruction of existing ones. In such cases the results of measurements, and the noise map, respectively, have a controlling role as analogous objects are used.

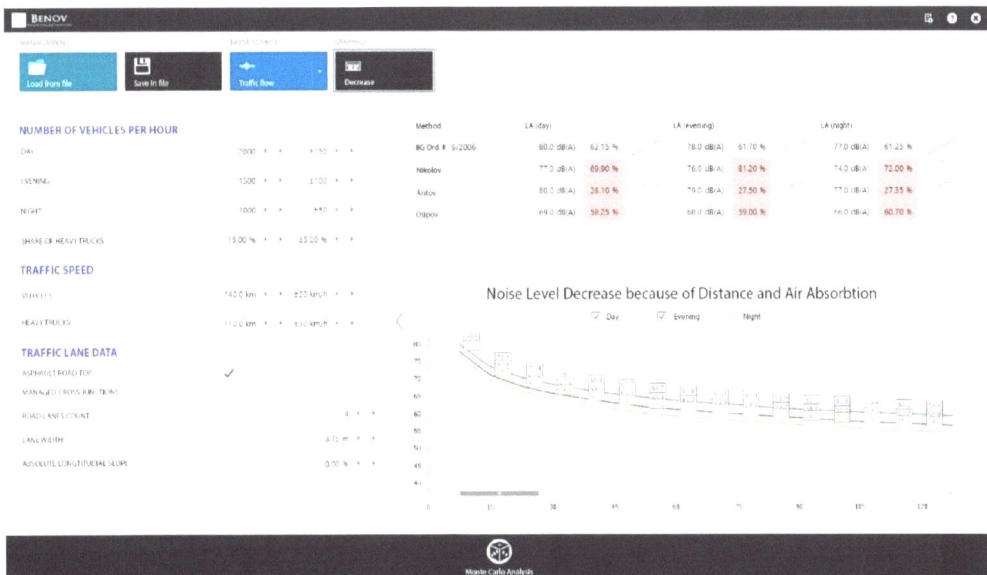

Figure X-3. Computing the noise level of traffic flow by using the Urban Acoustics software

The mathematical models underlying the analytical methods have been obtained by combining theoretical research with nature, and more rarely, laboratory observations.

Theoretical research points out the main factors determining the noise level of traffic flow:

- the intensity of the flow: N, expressed as a number of vehicles per one hour;

- the traffic speed for individual categories: V, km/h;
- the relative share of trucks: p, %;
- the corrections taking into account the type of road surface: asphalt, pavement, longitudinal slopes, etc..

According to N. Nikolov, D. Benov, I. L. Shubin (2014), the noise level of traffic flow in dB(A) at a distance of 7.5 m from the axis of the uttermost traffic lane is determined by the formula:

$$L_{Aeq}(7.5) = L_{Aeq}(25) + \Delta L_{Av} + \Delta L_{Asurf} + \Delta L_{Asl} + 6.95, \text{dB(A)}, \quad \text{(X.7)}$$

where

$$L_{Aeq}(25) = L_{Ab.eq}(25) + \Delta L_{Av} + \Delta L_{Asurf} + \Delta L_{Asl} - 1.23, \text{dB(A)}, \quad \text{(X.8)}$$

$L_{Aeq}(7.5)$ and $L_{Aeq}(25)$ are the noise level of traffic flow at the maximum permitted speed and asphalt concrete road surface.

ΔL_{Av} is a correction taking into account the influence of the maximum permitted speed;

ΔL_{Asurf} is a correction taking into account the influence of the road surface — for asphalt 0 dB(A), for asphalt concrete 2 dB(A), for pavement 3 dB(A);

ΔL_{Asl} is a correction taking into account the influence of the longitudinal slope: for a slope of 6% it is 0.6 dB(A), for 7% it is 1.2 dB(A), etc.

The base noise level $L_{Ab.eq}(25)$ in formulae (X.7) and (X.8) is determined by the formula:

$$L_{Ab.eq}(25) = 37.3 + 10\lg[N(1 + 0.082p)], \text{dB(A)}, \quad \text{(X.9)}$$

where N is the average hourly traffic intensity for the respective period of the day, veh/h;

p is a coefficient taking into account the number of trucks, as a percentage of total intensity; it is empirically established.

The correction for maximum permitted speed depending on the speed of motorcars and trucks: V_M and V_T, is determined by the formula

$$\Delta L_{Av} = L_M - 37.3 + 10\lg\left[\frac{(100+(100.1 L_{CV}-1)\rho)}{(100+8.23\rho)}\right], \text{dB(A)}, \qquad \text{(X.10)}$$

where $L_M = 27.7 + 10\lg[1 + 0.06 V_M]$;

$L_{CV} = L_M - 23.1 - 12.5\lg V_T$.

The approach adopted when building mathematical models presumes that an integral noise level of traffic flow is obtained, i.e. the possible local extremums are not taken into account.

Thus, we will examine two factors whose individual or joint effect determines the occurrence of significant deviations in the computed levels:

— the situation of traffic flows along the width of highway, in roadways A and B and lanes 1-4 (Figure X-4);

— dynamics of the indicators of intensity and speed.

For the application of the Monte Carlo method in input parameters are divided into constant and variable ones. Constant parameters are the number and width of lanes and the road surface of the roadway. Variable parameters are traffic intensity as well as the division of inner and outer lanes, traffic speeds of motorcars and trucks, the share of trucks and the longitudinal slope. Variable quantities can be set in different ways. We have assumed a uniform distribution, i.e. they are set in the form of $X \pm \Delta X$. If we assume that any of the variable quantities is constant, then we set $\Delta X = 0$.

1. *Influence of the situation of vehicles along the width of the road.* Figure X-4 shows a cross section of a highway with two roadways of 2X1 lanes each. The noise level is determined at the receiver, situated at a distance of 7.5 m from the axis of the uttermost traffic lane. This level, however, must be obtained by energetic summation of the levels of the flows from the four lanes as the distance between them is commensurable to the base distance (D. Benov, M. Mazhdrakov, N. Nikolov et al., 2013). Moreover, we must take into account the different conditions for traffic in

individual lanes. In inner lanes fewer cars travel but at higher speed. In outer lanes the speed is lower but the heavy vehicles travel there: trucks and buses.

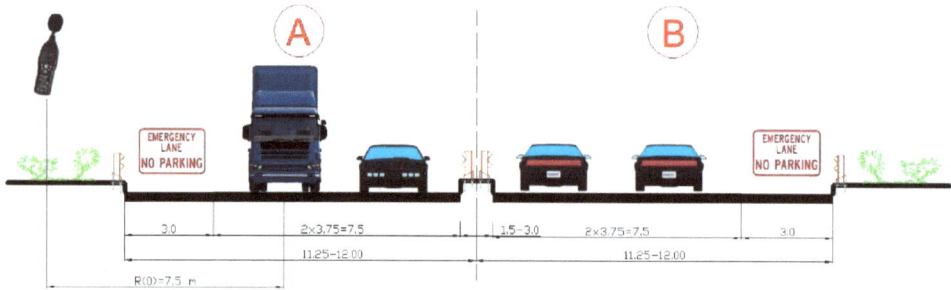

Figure X-4. Standard cross section of a highway

Due to the considerable width of the road the decrease in noise the farther one gets from the point of calculation must be taken into consideration by formula (X.5).

The resulting noise level is computed with energetic summation –

$$L = 10\lg \sum 10^{0.1L_i}, \mathrm{dB(A)}, \qquad (\mathrm{X.11})$$

where L_i is the noise level of the ith source, dB(A).

The situation of vehicles is taken into account by the ratio of the number of vehicles in roadways A and B, and in lanes 1 and 3, and 2 and 4, respectively. It should be taken into consideration that heavy vehicles travel on outer lanes and that the speed in inner lanes is higher.

2. Influence of the change in factors determining the noise level. We determine the noise level in different alternative situation of the vehicles on traffic lanes and roadways by the described method.

Figure X-5 shows the results of modeling in case of variation of factors:

- relative share of automobiles in roadway A: 0.50 ± 0.10;

- relative share of automobiles in the inner lane of roadways A and B: 0.70 ± 0.20;

- intensity of traffic flow: 2300 ± 500 automobiles/h;

- speed of motorcars in inner traffic lane: 100±30 km/h, and in the outer lane: 130±30 km/h;

- speed of trucks: 90 ± 10 km/h;

- percentage ratio of trucks: 30% ± 10%;

- longitudinal slope: 2‰ ± 2‰.

The following constants participate in the computations:

- number of traffic lanes: 2x2;

- width of one traffic lane: 3.75 m, and width of dividing line: 2 m;

- road surface type: asphalt.

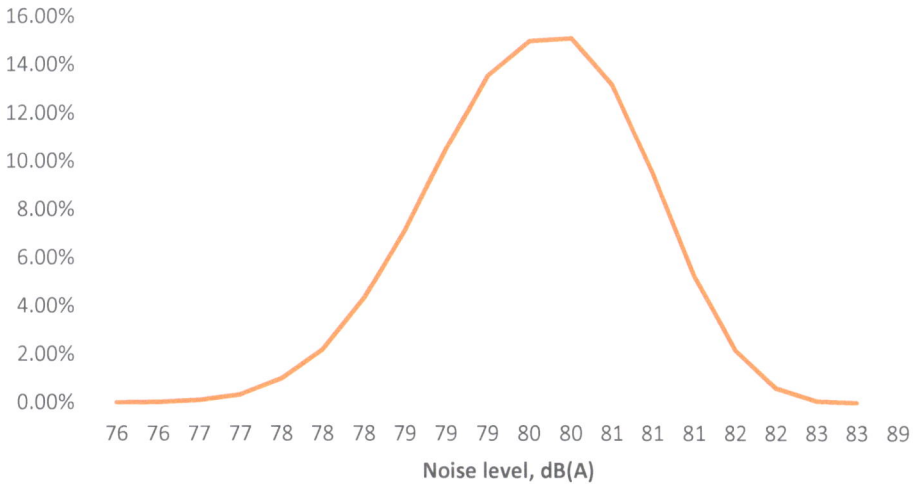

Figure X-5. Distribution of the noise level of traffic flow

3. *Local extremums of traffic flow intensity.* Extreme values are observed in traffic intensity, especially at entrance-exit highways of big cities, when the holiday time comes and on weekends, specially when the days off are "combined". The administrative measures

applied such as stopping heavy vehicle traffic, increased speed control and others have certain but limited effect on the noise load.

Observations show that in the afternoons of the last working day the outflows from big cities have an intensity of 4 to 5 thousand automobiles per hour in two lanes. Heavy vehicles is presented by limited numbers: 2 to 3%, buses in inner lane. Traffic speed in the two lanes is not higher than 80 to 110 km/h. The traffic in entrance lane is considerably weaker. Then, the noise level of the flow is about 80 dB(A), in the limits of 79 to 81 dB(A) (Figure X-5).

An analogous noise picture is observed in the entrance lane in the afternoons of the last non-working day.

In most countries with well-developed road infrastructure, the noise level is determined by means of formulae analogous to (X.7) -(X.10). For instance, in Switzerland (1995) the following formula is regulated:

$$L = 1 + 10 \log M +$$
$$+10 \log \left[\left(1 + \left(\frac{v}{50}\right)^3\right) \times \left(1 + B^* Eta \times \left(1 - \frac{v}{150}\right)\right) \right] \qquad (X.12)$$

where A and B are empirically determined parameters (according to the most recent measurements $A = 43$ and $B = 20$);

v is the traffic speed, km/h;

Eta is the share of trucks;

M is traffic intensity, veh/h.

When compared with the formulae adopted in Bulgaria the following can be observed:

- the considerable influence of speed;

- the influence of road conditions (type of road surface, slopes, etc.) is taken into account integrally with the parameter A.

Traffic intensity is determined by counting the vehicles (E. Schweiz, 2014). The highest intensity is along highway A1: 144,134 automobiles in the Canton of Zürich and 129,932 automobiles in the

Canton of Basel-Country. The hourly intensity is determined by the formulae

- $N_{day} = 0.058 \times M$;
- $N_{night} = 0.009 \times M$.

For example, for the Canton of Zürich the daytime hourly intensity is 8360 automobiles per hour and the night-time is 1169.

The standard cross section of highways in Switzerland is practically the same as the one adopted in Bulgaria (Figure X-4).

A simulation under the Monte Carlo method has been with the following data:

— traffic speed from 70 to 90 km/h;

— traffic intensity as per the data for the Canton of Zürich with a deviation of 10%;

— share of truck transport from 20 to 30 %.

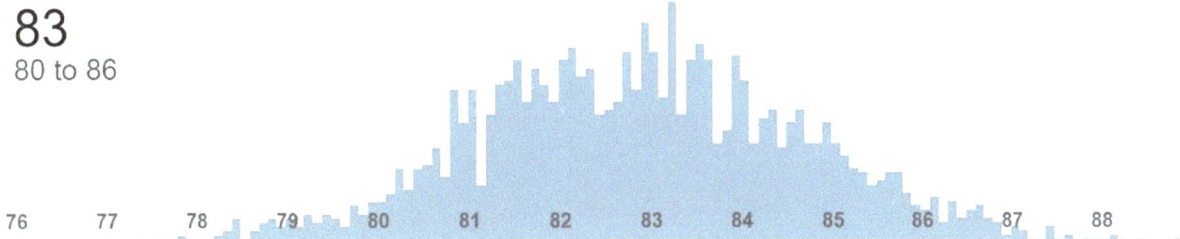

83
80 to 86

76   77   78   79   80   81   82   83   84   85   86   87   88

Figure X-6. Results from the application of the Monte Carlo method

The obtained results are shown in Figure X-6.

X.4. ESTIMATING THE NOISE LEVELS OF OPEN-AIR PUBLIC CARPARKS

One of the most difficult transport & urban planning problems in present-day cities is parking planning and arrangements. The difficulty comes from the dynamism in time and space of the quantitative parameters of traffic's: the number of automobiles, duration of stay, situation within the populated area, etc. These problems are quite complex for central urban areas where various public functions are concentrated, flows of motor vehicles are intensive and the need of parking spaces is exceptionally great (N. Nikolov, A. Kovachev, D. Benov et al., in print).

A crucial problem is not only to ensure areas but also to find their appropriate location in order to ensure both the safety of urban traffic and the reaching of environmental standards. High intensity of building development typical for all mid-sized and big cities creates further difficulties for construction of the necessary carparks and garages in the downtown urban core.

Parking planning in the Republic of Bulgaria is made at a level of motorization of 320 personal automobiles per 1,000 citizens for the stage until 2010. Currently, this indicator, however, has been considerably exceeded. Mass transport vehicles, trucks and taxis, state-owned automobiles are excluded when determining the level of motorization. Areas for depots, garages, repair stations, service stations are needed for the mass urban passenger transport (bus, trolley bus, tramway and others).

A standard of one parking space per each residence has been adopted for dimensioning of places for parking in residential territories. No more than 50% of the number of parking spaces determined by statute is envisaged for outdoor parking in the immediate vicinity of buildings. The remaining portion of parking places is designed in underground, semi-underground and overground storied carparks and garages that may be situated outside the area of the respective structural element.

Mostly open-air carparks are envisaged manufacturing and trade areas. A parking plan is developed for populated areas of zero and 1-st functional type (at purposefulness and at IInd functional type) for central structural elements. The plan is used to determine the location, capacity and stages of construction of carparks. In the case if such populated areas different systems conforming to the system of mass passenger transport are used in the actual urban planning practice: the "P+T" system (park and travel), time-limitation of parking, introduction of paid parking ("blue zone") and others.

The parking standards according to the requirements on planning and designing the communication and transportation systems of populated areas (N. Nikolov, D. Benov, I. L. Shubin, 2014), are shown in Table X-2.

It is accepted that when determining the number of parking spaces on the basis of dynamic standards the following adjustment coefficients should apply: 1.3 for the central urban core; 1.1 for the central urban part; 1.2 for regional centers. For all other urban areas this coefficient is 1.0.

Table X-2. Parking rules

No	Site	Unit	One space of
1	Administrative and office buildings	gross floor area	40 to 60 m²
2	Big shopping centers and stores (MALL, chain stores)	gross floor area	100 to 120 m²
3	Covered markets	gross floor area	150 to 200 m²
4	Cinemas, universal halls	seats	10 to 20 m²
5	Representative hotels (more than ****)	beds	3 to 5 pcs
6	Restaurants – café	chairs	8 to 10 pcs
7	Universities	staff students	3 to 5 persons 15 to 25 persons
8	Sport halls, covered swimming pools	visitors	10 to 15 persons
9	Stadiums, open sports playgrounds	visitors	30 to 40 persons
10	Railway stations and bus stations	passengers/hour	15 to 20 persons
11	Hospitals and sanatoriums	beds	8 to 12 pcs.

Carparks are classified by several indicators.

1. By capacity carparks are divided in five categories and according to the statutory rules in force (N. Nikolov, D. Benov, I. L.

Shubin, 2014) they are situated at a minimum distance from residential buildings as per Table X-3.

For some public buildings (schools, kindergartens) the permitted distances are greater, such as multi-storied carparks for motorcars with capacity of more than 100 parking spaces, a sanitary protection area of 50 m is determined (N. Nikolov, A. Kovachev, D. Benov et al., in print).

Table X-3. Categories of carparks

Category	Number of parking spaces, pcs.	Minimum distance, m
I	> 100	25
II	51 to 100	20
III	26 to 50	15
IV	11 to 25	10
V	3 to 10	10

2. By location, parking can be made in:

- on-street parking areas: on street lane (outside the lane for traffic flow);

- off-street parking areas: open-air public carparks.

In off-street carparks there might be underground, semi-underground and overground parking. Carparks are overground where the level of the pavement is the same or above the level of the adjacent terrain. In-door carparks can be single- or multi-storied.

Carparks must be constructed to each residential, public or manufacturing site subject to the architectural and urban planning, sanitary and hygiene and fire safety requirements.

Figure X-7 shows a carpark on a site of a trade chain store with 220 parking spaces and plan composition at an angle of 90.

The approximate averaged areas for parking per one vehicle together with the area needed for maneuvers (to get to and leave the space) are shown in Table X-4.

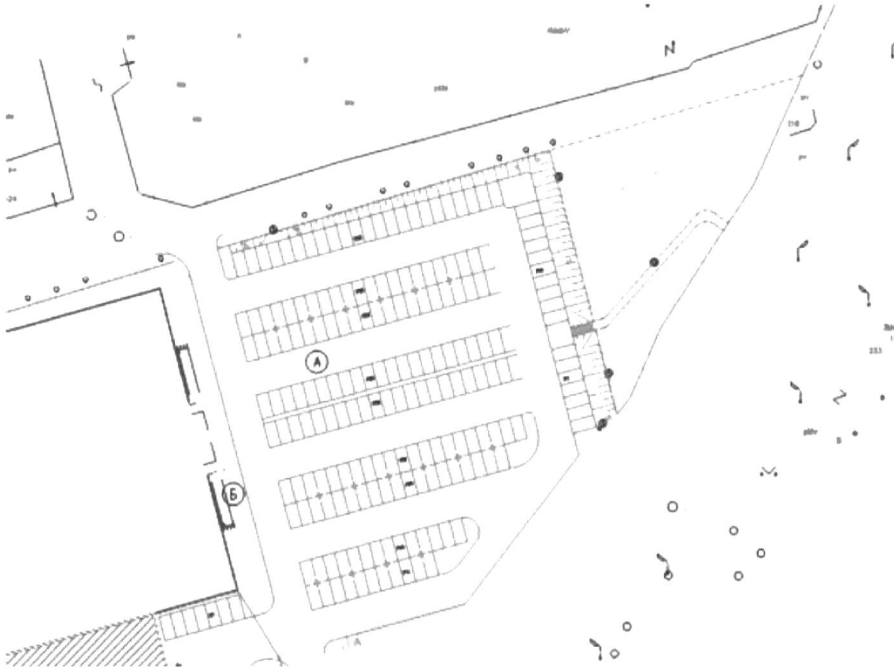

Figure X-7. Carpark of a chain store

Table X-4. Approximate averaged areas for parking per one vehicle

Vehicle type	Area, m^2
motorcar	25
truck	30-60
bus (up to 24 seats)	up to 40
bus (more than 24 seats)	up to 50
bus (articulated)	up to 65
motorcycle	up to 4

Overground carparks are the most crucial noise source and their location must be in conformity with a number of requirements related to the impact of noise, vibrations and distance from residential buildings.

Carparks are defined as local noise sources in urbanized areas and the equivalent noise levels in the area of the impact places

L_{Aterr}, dB(A) per day, evening and night (period t=12,4,8 h) are determined by the formula

$$L_{Aterr} = L_{Aeq}(25) - \Delta L_{Ad} - \Delta L_{Abarr}, \text{dB(A)}, \qquad (\text{X.13})$$

where $L_{Aeq}(25)$ is the noise level of the source, dB(A);

ΔL_{Ad} is the decrease in noise level, dB(A), depending on the distance between the source and the point of computation;

ΔL_{Abarr} is the decrease in noise level, dB(A), from noise protection facilities such as noise protection embankments and screens, relief, green plantations, etc.

The noise level $L_{Aeq}(25)$, dB(A), for public carparks at a distance of 25 m from the boundary of the source is calculated by the formula

$$L_{Aeq}(25) = 37 + 10 \lg(Nn) + \Delta L_{Ac} - 1.23, \text{dB(A)}, \quad (\text{X.14})$$

where n is the number of parking spaces;

N is the average number of motor vehicles having entered and exited one parking space for certain period of the day (averaged per 1 h);

ΔL_{Ac} is a correction taking into account the different types of vehicles; for carparks for automobiles $\Delta L_{Ac} = 0$ dB(A), for trucks: $\Delta L_{Ac} = 10$ dB(A), and for motorcycles: $\Delta L_{Ac} = 5$ dB(A)

When determining the noise level of carparks the parameter N for the computation of which there is no formula apparatus is present in the second term of the formula (X.14). This parameter can be determined by processing of statistical data obtained from counting after the carpark is constructed. This, however, does not allow sufficiently accurate computations to be made in design phase to determine the necessary sanitary and protection area to the line of building development of the residential area that borders on it.

For estimation of the acoustic impact of open-air overground carparks, a probabilistic model of the "parking" process has been prepared based on the theory of mass service (E. S. Wentzel, 1982).

When preparing the model two factors have been taken into consideration: users and parking regulation.

Users are divided into separate groups on the basis of their behavior. An example of the conduct of individual groups of users in case of a public carpark with capacity of 100 parking spaces is presented in Table X-5. For carparks to trade outlets, business parks, etc. the distribution of the groups and their behavior will differ. For the quantitative expression of the behavior of each group the expected (estimated) minimum and maximum values of the following indicators are used:

- number of automobiles;
- time of arrival in the carpark;
- duration of stay.

The organization of parking is reflected by means of:

Table X-5. Example of the behavior of different groups of users of public carpark with a capacity of 1000 places

Type of user	Number		Arrival time		Stay, h	
	from	to	from	to	from	to
Lessees	0	20	6	6	24	24
Office lessees	5	10	8	10	8	12
Shopper	200	700	9	20	0.3	2
Eating lunch at restaurant	5	20	12	14	1	3
Eating dinner at restaurant	1	30	19	22	2	4
Visitors of night clubs	5	20	22	24	4	6
Visitors of cinemas, theaters	20	50	16	22	2	4
Visitors of offices	10	20	8	16	0.5	2
Casual, day-time	20	40	7	20	0.5	2
Casual, night-time	10	20	20	2	0.5	2
Staying at the hotel	5	10	14	20	8	24

- the effective number of parking spaces (which is not always the same as the designed number of spaces);

- the presence of "booked" spaces which are kept for certain user only;

- the possibility of a "queue" of waiting automobiles.

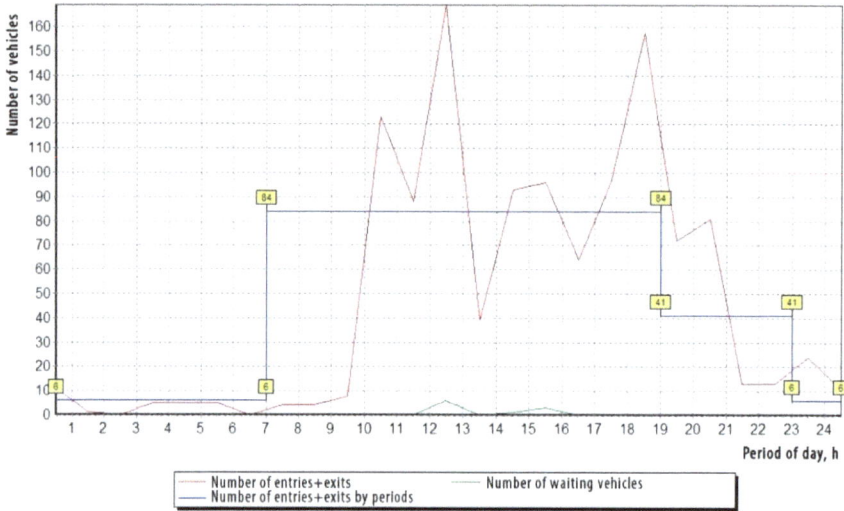

Figure X-8. Entries and exits during a 24-hour period into/from a carpark with capacity of 1,000 spaces

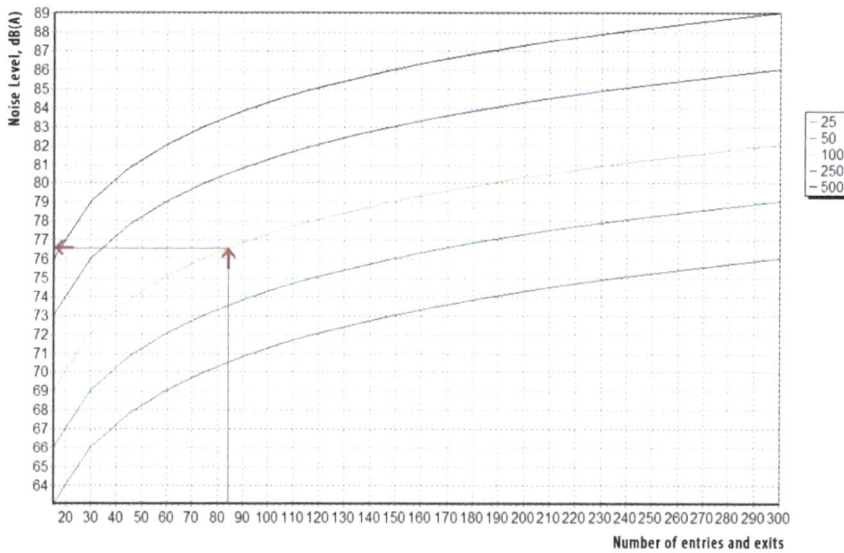

Figure X-9. Graphs of changes in the noise level at different carpark capacities

Urban Acoustics program realizes a stochastic model of the functioning of an open-air public carpark. Graphs with hourly values of two groups of users are obtained as a result of program's operation. The first group characterizes the carpark as an investment site; these are the maximum, average and minimum numbers of automobiles parked. The second group characterizes the carpark as a mass service system; this is the sum of entries and exits ("intensity of requests") and the number of waiting automobiles ("queue length").

Figure X-8 shows the result of a run of the program for carpark for public use with capacity of 100 parking spaces. In individual time zones the sum of entries and exits is 84, 41 and 6, respectively.

The noise level of carparks is computed by a module of Urban Acoustics program, which is made on the basis of the formula (X.14), as the value of the parameter N is estimated by the aforesaid manner. The base input information includes the number of parking spaces and the sum of entries and exits into/from the carpark.

Figure X-9 shows the graphs of change in the noise level of carparks with capacity of 25, 50, 100, 250 and 500 parking spaces in 15 to 300 entries and exits. For example, for a carpark with capacity of 100 parking spaces and average number of entries and exits for the period from 7h to 19h – 84, the noise level is 76.5 dB(A).

From the stochastic studies made it can be concluded that the sum of entries and exits of automobiles into/from the carpark, which is a decisive parameter in the formula (X.14) (N. Nikolov, D. Benov, I. L. Shubin, 2014), depends, in the first place, on the structure of users and differs depending on the intended use of the carparks.

X.5. MODELING OF RAILWAY NOISE USING MONTE CARLO METHOD

In building acoustics is accepted that for noise level of the railway flows should be used the equivalent sound level L_{Aeq} and maximum noise level L_{Amax} в dB(A) at basic distance 25 m from the track axis.

According to the regulated "Method for calciulating the noise from rail traffic" (N. Nikolov, D. Benov, I. L. Shubin, 2014), the noise level of railway flows $L_{Aeq}(25)$, dB(A), passing by rail road (track) for the period during the day, in the evening or during the night period at a maximum speed of 100 km/h and an upper ballast prism construction and grate of wooden sleepers is determined by the formula:

$$L_{Aeq}(25) = 51 + 10\lg[N(5 - 0.04p)], dB(A), \quad (X.15)$$

where N is the average number of train compositions from one type per hour; It is determined by the number of trains of the respective type for the assessment period;

p – the number of wagons with disc brakes (in%) in the train composition.

The equivalent sound level at the calculation points of the territory L_{Aterr}, dB(A), is determined by the formula:

$$L_{Aterr} = L_{Aeq}(25) + \Delta L_A + \Delta L_{Al,v} + \Delta L_{Aup.c} -$$

$$-(\Delta L_{Ad} + \Delta L_{Ar} + \Delta L_{Abarr} + \Delta L_{Asub}), dB(A), \quad (X.16)$$

where ΔL_{Awag} is a correction, considering the type of wagons, dB(A); for wagons with disc brakes -2 dB(A), for freight wagons it is 0 dB(A);

$\Delta L_{Al,v}$ – a correction, considering the lenght ℓ, m, of the rain composition and the speed V, km/h..

$$\Delta L_{Al,v} = 10\lg(\ell V) - 60, dB(A). \quad (X.17)$$

In case of parameters ℓ and V are not known, their values are used from Table X-6 (N. Nikolov, D. Benov, I. L. Shubin, 2014).

Table X-6. Parameters of train compositions

Type of composition	Maximum speed V, km/h	Average lenght ℓ, m
Passenger train (express, intercity)	100	270 to 300
Freight train	100	500

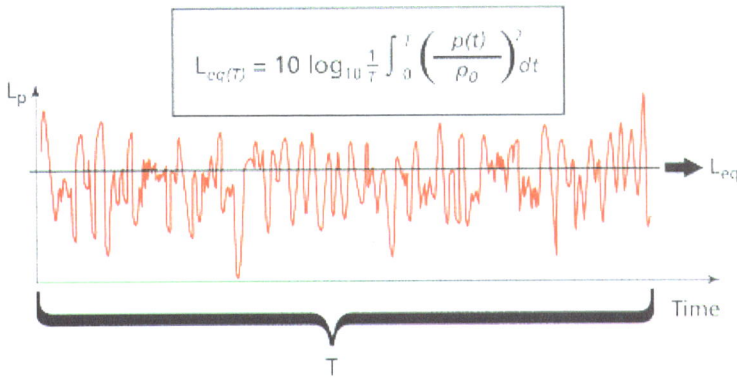

$$L_{eq(T)} = 10 \log_{10} \frac{1}{T} \int_0^T \left(\frac{p(t)}{p_0} \right)^2 dt$$

Figure X-10. Dependence between L_{Aeq} and L_{Amax}

$\Delta L_{Aup.c}$ – a correction, considering the different types of upper construction of the rail track, dB(A); for a base ballast prism and a wooden sleepers grill $\Delta L_{Aup.c} = 0$ dB(A), for a base ballast prism and grill of concrete sleepers $\Delta L_{Aup.c} = +2 \, dB(A)$.

ΔL_{Ad} – the reduction of the sound level depending on the distance r to the calculation point, which is determined by the formula (X.15).

ΔL_{Ar} – the reduction of sound level due to the influence of the earth's surface and the atmospheric conditions, which is calculated by formula (X.16).

ΔL_{Asub} – a correction of +5 dB (A) for the irritating influence of rail noise.

ΔL_{Abarr} – reduction of noise levels in the presence of noise barriers.

Determining of noise level of railway flows is very important, because based on this is made on one hand evaluation of the expected noise level in the urbanized territories, and on the other hand is determined the choice of optimal noise protection measures (N. Nikolov, D. Benov, I. L. Shubin, 2014).

At speeds of up to 50 km/h, as a rule, the highest noise is emitted by the engine, in the range from 50 to 200 km/h, the noise generated by the interaction of the rails and wheels and at speeds above 200 km/h the aerodynamic noise day the greatest impact. (Figure X -11) (D. Benov, N. Nikolov, M. Mazhdrakov, 2017).

Based on the formulas, we have developed a probable model using the Monte Carlo method.

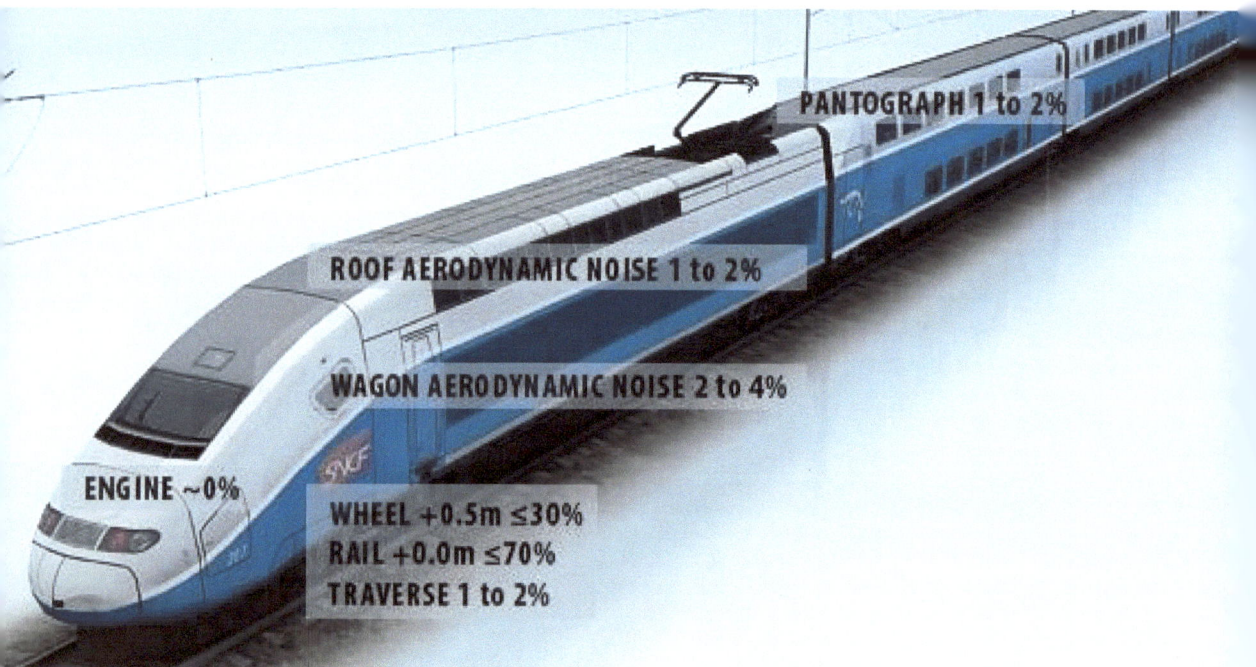

Figure X-11. Distribution of noise emissions at a speed of 50 to 200 km/h

In railways, many trains run on the same rails. To ensure fast enough, and at the same time save drive movement, the random element involved in the Monte Carlo model must be kept to a minimum. In practice, this is done with organizational and technological solutions and traffic management under a strict algorithm.

Therefore, when modeling, the parameters changes in narrow limits, this is especially true for the movement between the stations, when the train compositions are moving with practically constant speed (D. Benov, N. Nikolov, M. Mazhdrakov, 2017).

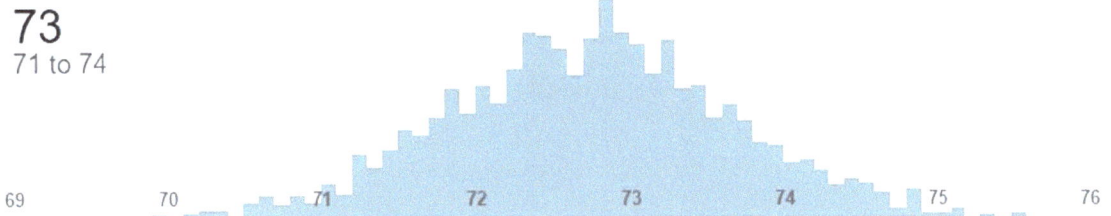

73
71 to 74

Figure X-12. Study results at a speed of 110 to 140 km/h

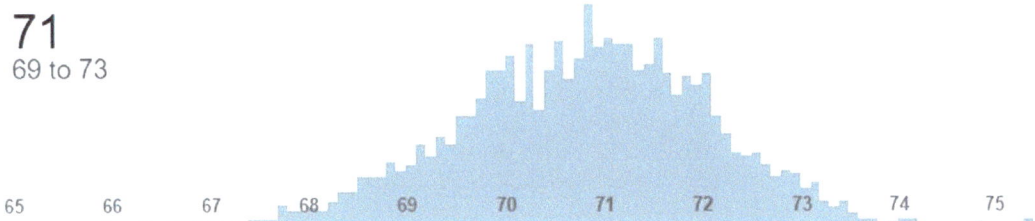

71
69 to 73

Figure X-13. Study results at a speed of 80 to 120 km/h

The input parameters of Monte Carlo model are:

– freight trains only, daily period (12h);

– 24 to 26 train intensity, evenly distributed;

– the lenght of trains – 300 to 500 m, evenly distributed;

– speed – 110 to 140 km/h (Figure X-12) and 80 to 120 km/h (Figure X-13), normally distributed.

The results are shown Figure X-12 and Figure X-13. As noted, at railway transport significant number of constraints are required. As a result, the resulting set of variants is within relatively narrow limits. This feature is clearly visible by comparing the results of D. Benov, M. Mazhdrakov, N. Nikolov et al. (2013), with those shown in Figure X-12 and Figure X-13.

X.6.   LOCAL NOISE SOURCES IN URBANIZED AREAS

Local noise sources in urbanized areas are construction sites, open-pit mines/quarries, industrial sites, etc.

As a rule, construction sites are situated in residential areas and the noise from construction and assembly works, due to its specific acoustic characteristics and duration causes serious discomfort among the residents of neighboring buildings.

Different construction machines, units and tools are used simultaneously at each individual technological stage upon construction of the sites. Works classified as "very noisy" such as concrete breaking and tearing, explosion works, cutting and milling, sand blasting, and others suchlike require special attention. The levels of noise emitted by construction machines at a distance of 7.5 m from the boundaries of construction site reaches 75 to 85 dB(A), and the noise from pile drivers reaches 100 dB(A) (N. Nikolov, D. Benov, 2010).

In order to reduce the sound power of construction machines technical standards have been developed in the European Union limiting the permitted levels of noise emitted by the specific machine. Designers and manufacturers must conform to these standards. For some types of machines the adjusted sound power level L_{W_A}, dB(A),is adopted as a statutory characteristic as such level is determined by the formula:

$$L_{W_A} = L_A + 10\lg\frac{S}{S_0}, \text{dB(A)}, \qquad (X.18)$$

where L_A is the noise level of the respective machine, dB(A);

S is the area of measurement surface situated at a distance R form the center of the machine to the point of computation determined by Table X-7; in a special case $S = 2\pi R^2$, m^2;

$S_0 = 1$ m^2.

In the formula (X.18) the measurement surface is a semishere, whose radius R depends on the base lenght ℓ, m, (Table X-7, Figure X-14).

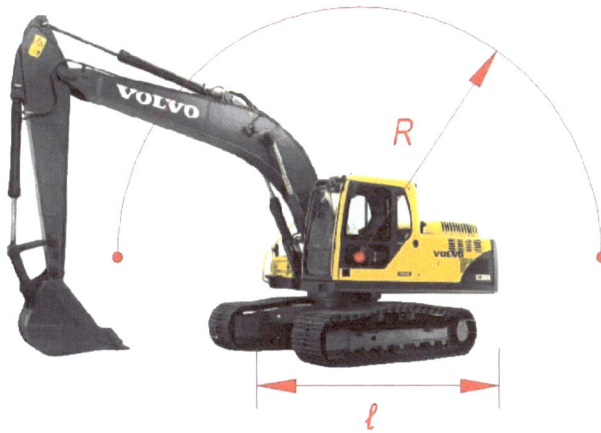

Figure X-14 Scheme of base length of construction machines and radius of measurement semisphere

Table X-7. Radius of measurement semisphere in different base length of construction machines

Base length of construction machine ℓ, m	Radius of measurement semisphere R, m
$\ell < 1.5$	4
$1.5 < \ell < 4.0$	10
$\ell \geqslant 4.0$	16

Table X-8. Technical standards for noise for construction equipment

Machine Type	Power N, kW	Rate of corrected level of sound power L_{W_A}, dB(A)
Compaction machines (steam rollers, shock vibrators)	$N \leqslant 8$	105
	$8 < N \leqslant 70$	106
	$N > 70$	$86 + 11\lg N$
Tracked (front tracked loaders, bulldozers, excavators)	$N \leqslant 55$	103
	$N > 55$	$84 + 11\lg N$
Wheeled (front loaders, bulldozers, motor graders, cranes)	$N \leqslant 55$	101
	$N > 55$	$85 + 11\lg N$
Compressors	$N \leqslant 50$	94
	$N > 50$	$95 + 11\lg N$
Wheeled excavators	$N \leqslant 50$	93
	$N > 50$	$80 + 11\lg N$

The technical standards on the noise emitted by construction machines depend on the type of machine and the power of the engine (Table X-8).

From Table X-8 it is evident that the difference in the standard parameters for different machines reaches 12 dB(A).

According to N. Nikolov, D. Benov, I. L. Shubin (2014), the equivalent levels of noise L_{Aterr}, dB(A), from local noise sources at the point of computation during the day, in the evening, and at night (period T = 12,4,8 h), are determined by the formula:

$$L_{Aterr} = L_{Aeq}(r_0) - \Delta L_{Ad} - L_{Abarr}, \text{dB(A)}, \qquad (X.19)$$

where $L_{Aeq}(r_0)$ is the noise level of the source determined at a base distance $r_0 = 7.5$ m, dB(A);

ΔL_{Ad} is the decrease in noise level depending on the distance between the noise source and the point of computation, dB(A);

L_{Abarr} is the decrease in noise level due to screening facilities (noise barriers), dB(A).

As stated above the noise level of construction sites depend on the type of construction and assembly works being performed, some of which are given in (Table X-9).

Table X-9. Noise emitted by some types of Construction and Assembly Works

Type of Construction and Assembly Works	Noise level $L_A(7.5)$, dB(A)
Loading-unloading	71
Excavation	77
Asphalt laying	80
Asphalt cutting	85
Pile driving	90

Research by (N. Nikolov, D. M. Benov, M. Mazhdrakov, 2016) shows that in case of simultaneous operation of two and more construction machines and/or transport vehicles the construction site can be considered a source of limited length of quasi-cylindrical sound waves. Then, the decrease in noise level in case of each doubling of distance is in the range from 3 to 6 dB(A). The specific value of the noise level depends on the number of sources and the distance between them and is variable depending on the distance between the source and the point of computation (Figure X-15, Table X-10).

The values of summary and relative decrease in the noise level if doubling the distance are given in Table X-10.

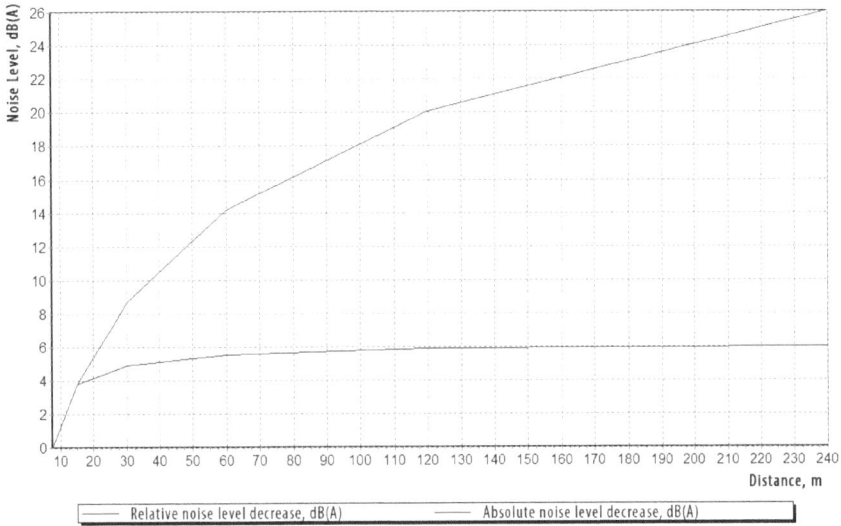

Figure X-15. Decrease in the noise levels depending on the distance r

Table X-10. Summary and relative decrease in the noise level if doubling the distance

r	15	30	60	120	240
ΔL_{Ad}	4.0	9.1	14.8	20.8	26.8
$\Delta L_A(r) - \Delta L_A(2r)$	4.0	5.1	5.7	6.0	6.0

The decrease in noise levels due to the distance depending on the number of individual sources n and the distance between them ℓ, ΔL_{Ad} is calculated by one the following formulae (Figure X-16):

– in case of two individual sources (n=2):

$$\Delta L_{Ad} = 20 \lg \frac{\sqrt{(2r)^2 + \left(\frac{\ell}{2}\right)^2}}{\sqrt{r^2 + \left(\frac{\ell}{2}\right)^2}} ; \qquad (X.20)$$

– in case of three or four individual sources (n=3,4):

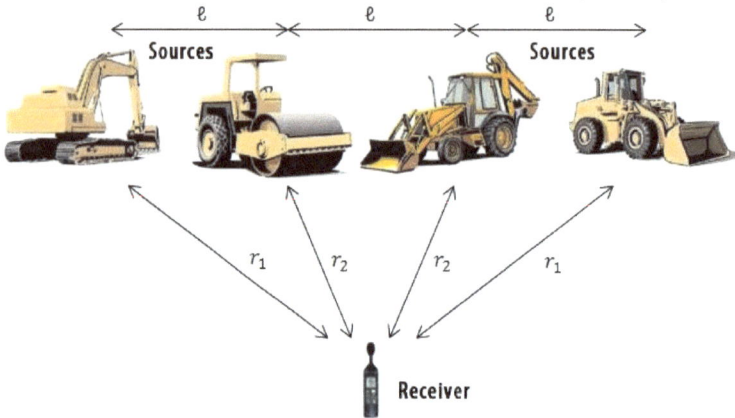

Figure X-16. Scheme of computing the decrease in noise levels at n=4.

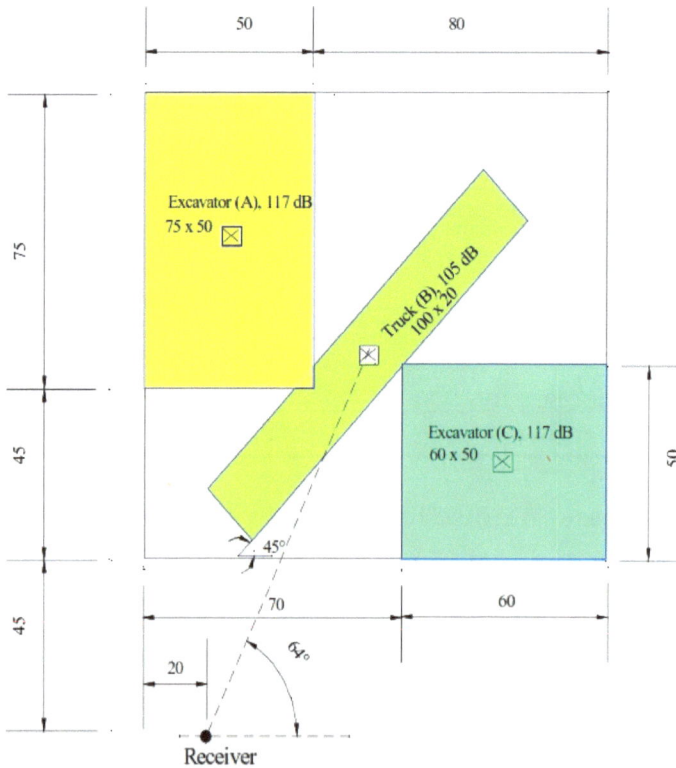

Figure X-17. Configuration of construction site

$$\Delta L_{Ad} = 20 \lg \frac{\delta_{r-1}}{\delta_r}, \text{dB(A)}; \qquad (X.21)$$

where δ_r is a parameter that depends on the number of individual sources and the distances from them to the point of computation:

– in case of n=3:

$$\delta_r = \sqrt{\frac{4}{r_1^2} + \frac{1}{r_2^2} + \frac{4}{r_1 r_2}}, \text{m}; \qquad (X.22)$$

– in case of n=4:

$$\delta_r = \sqrt{\frac{1}{r_1^2} + \frac{1}{r_2^2} + \frac{1}{r_1 r_2}}, \text{m}. \qquad (X.23)$$

Most often more than 2 to 3 machines are operating on the construction site. Some of them are stationary and the others move in certain perimeter, including outside the site. Then, the Monte Carlo method is appropriate for modeling of the noise level of construction sites because it reflects the dynamics and the complex interaction of individual noise sources.

Figure X-17 shows the operating perimeters of 2 excavators and one dump truck on a construction site (Z. Haron, K. Yahya, 2009).

The results from noise level modeling of the site under consideration by the Monte Carlo method are given in Figure X-18 (N. Nikolov, D. Benov, M. Mazhdrakov, 2017).

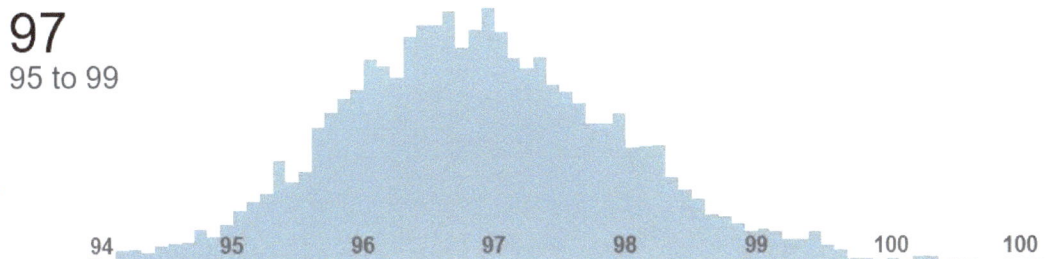

Figure X-18. Noise level of construction site and standard deviation

The obtained symmetrical distribution corresponds to the technological scheme (Figure X-17), as the nearest source is the one having the greatest influence.

REFERENCES

[1] Benov, D., M. Mazhdrakov, N. Nikolov, et al. (2013). Detailed modeling of traffic noise on highways. VIth Russian Scientific-Practical Conference with International Participation "PROTECTION FROM HIGH NOISE & VIBRATION". N. I. Ivanov. Saint Petersburg, Baltic State Technical University "Voenmeh": 477-482 [in Russian].

[2] Benov, D., N. Nikolov, M. Mazhdrakov (2017). Modeling of railway noise using Monte Carlo method. International Conference "Problems of ecological safety, energy saving in construction and housing and communal services". Moscow-Kavala: 17-21 [in Russian].

[3] Haron, Z., K. Yahya (2009). "Monte Carlo Analysis for Prediction of Noise from a Construction Site." Journal of Construction in Developing Countries 14(1): 1-14.

[4] Landau, L. D., E. M. Lifshitz (1987). Fluid Mechanics, Butterworth-Heinenann.

[5] Nikolov, N., D. Benov (2010). Local noise sources in urbanized areas. Scientific Conference with International Participation "Science and Society", Kardzhali, Bulgaria [in Bulgarian].

[6] Nikolov, N., D. Benov, M. Mazhdrakov (2017). Modeling of local noise sources in residential areas using Monte Carlo method. International Conference "Problems of ecological safety, energy saving in construction and housing and communal services". Moscow-Kavala: 86-91 [in Russian].

[7] Nikolov, N., D. Benov, I. L. Shubin (2014). Acoustic Design of Transport Noise Barriers. Sofia, Bulgaria, ACMO Academic Press [in Bulgarian].

[8] Nikolov, N., D. M. Benov, M. Mazhdrakov (2016). Application of the theory of quasicylindrical waves in acoustic calculations. Saarbrucken, LAP Lambert Academic Publishing [in Russian].

[9] Nikolov, N., A. Kovachev, D. Benov, et al. (in print). <u>Urban Planning Acoustics</u>. Sofia, ACMO Academic Press [in Bulgarian].

[10] Osipov, G. (2004). <u>Sound insulation and sound absorption</u>. Moscow, Astrel [in Russian].

[11] Schweiz, E. (1995). Korrekturen zum Strassenlärm-Berechnungsmodell. W. u. L. B. Bundesamt für Umwelt. Bern.

[12] Schweiz, E. (2014). Messstellen mit dem höchsten durchschnittlichen Tagesverkehr, 2014. A. S. a. S. (SASVZ). Bern.

[13] Wentzel, E. S. (1982). <u>Probability Theory (first Steps)</u>, Mir Publishers.

UNCLE PETYO WAS RIGHT:

THERE IS SOMETHING HERE!

www.ingramcontent.com/pod-product-compliance
Lightning Source LLC
Chambersburg PA
CBHW050826220326
41598CB00006B/324